Ute Henniges

Cellulose Analytics by Gel Permeation Chromatography

Ute Henniges

Cellulose Analytics by Gel Permeation Chromatography

Description of degradation processes, evaluation of stabilisation procedures and development of a non-destructive alternative

Südwestdeutscher Verlag für Hochschulschriften

Impressum/Imprint (nur für Deutschland/ only for Germany)
Bibliografische Information der Deutschen Nationalbibliothek: Die Deutsche Nationalbibliothek verzeichnet diese Publikation in der Deutschen Nationalbibliografie; detaillierte bibliografische Daten sind im Internet über http://dnb.d-nb.de abrufbar.
Alle in diesem Buch genannten Marken und Produktnamen unterliegen warenzeichen-, marken- oder patentrechtlichem Schutz bzw. sind Warenzeichen oder eingetragene Warenzeichen der jeweiligen Inhaber. Die Wiedergabe von Marken, Produktnamen, Gebrauchsnamen, Handelsnamen, Warenbezeichnungen u.s.w. in diesem Werk berechtigt auch ohne besondere Kennzeichnung nicht zu der Annahme, dass solche Namen im Sinne der Warenzeichen- und Markenschutzgesetzgebung als frei zu betrachten wären und daher von jedermann benutzt werden dürften.

Verlag: Südwestdeutscher Verlag für Hochschulschriften Aktiengesellschaft & Co. KG
Dudweiler Landstr. 99, 66123 Saarbrücken, Deutschland
Telefon +49 681 37 20 271-1, Telefax +49 681 37 20 271-0, Email: info@svh-verlag.de
Zugl.: Wien, BOKU, Diss., 2008

Herstellung in Deutschland:
Schaltungsdienst Lange o.H.G., Berlin
Books on Demand GmbH, Norderstedt
Reha GmbH, Saarbrücken
Amazon Distribution GmbH, Leipzig
ISBN: 978-3-8381-0437-9

Imprint (only for USA, GB)
Bibliographic information published by the Deutsche Nationalbibliothek: The Deutsche Nationalbibliothek lists this publication in the Deutsche Nationalbibliografie; detailed bibliographic data are available in the Internet at http://dnb.d-nb.de.
Any brand names and product names mentioned in this book are subject to trademark, brand or patent protection and are trademarks or registered trademarks of their respective holders. The use of brand names, product names, common names, trade names, product descriptions etc. even without a particular marking in this works is in no way to be construed to mean that such names may be regarded as unrestricted in respect of trademark and brand protection legislation and could thus be used by anyone.

Publisher:
Südwestdeutscher Verlag für Hochschulschriften Aktiengesellschaft & Co. KG
Dudweiler Landstr. 99, 66123 Saarbrücken, Germany
Phone +49 681 37 20 271-1, Fax +49 681 37 20 271-0, Email: info@svh-verlag.de

Copyright © 2009 by the author and Südwestdeutscher Verlag für Hochschulschriften Aktiengesellschaft & Co. KG and licensors
All rights reserved. Saarbrücken 2009

Printed in the U.S.A.
Printed in the U.K. by (see last page)
ISBN: 978-3-8381-0437-9

I'm on the road again
and I'm searching for
the philosopher's stone.

Van Morrison, 1999: Back on Top

Acknowledgement

Mein besonderer Dank gilt Ao.Univ.Prof. Dipl.-Chem. Dr.rer.nat Antje Potthast für die aufmerksame Betreuung dieser Arbeit und die stetige, sehr engagierte und persönliche Unterstützung. Gleichermaßen möchte ich mich bei Prof. Dipl. Ing. Dr. Gerhard Banik bedanken, der diese Arbeit wesentlich mitbetreut hat und zahllose Probenmaterialien zur Verfügung gestellt hat. Univ.Prof. Dr.rer.nat. Dipl.-Chem. Thomas Rosenau danke ich herzlich für vielerlei Anregungen. Bei Univ.Prof. Dipl.-Ing. Dr. Paul Kosma möchte ich mich für die Möglichkeit bedanken, an seinem Department zu promovieren.

Bei Dr.nat.techn. Dipl.-Ing. Sonja Schiehser bedanke ich mich für die tatkräftige Unterstützung im Labor und bei den Auswertungen, sowie für die gute Zusammenarbeit. Gleiches gilt für Dr.nat.techn. Dipl.-Ing. Manfred Schwanninger, der sich zusätzlich um die technische Unterstützung bei der Durchführung eines Teiles dieser Arbeit gekümmert hat und von Frau Dr. Dietl (Firma Bruker) immer wieder das MPA ausgeliehen hat, wofür ihr an dieser Stelle ebenfalls gedankt sei.

Maria Hobel danke ich für ihre Geduld bei eiligen Chemikalien- und Glaswarenbestellungen, sowie bei fehlerhaft ausgestellten Anträgen, von denen ich eine Menge produziert habe. Einen Großteil des von mir verbrauchten Fluoreszenzmarkers hat Andreas Pitschmann hergestellt, der mich darüber hinaus auch teilweise in ihre Herstellung eingeweiht hat. Bei Dr. Mirjana Kostic und Dr. Gentiana Nagel bedanke ich mich für einen sehr guten Start an der BOKU und an der GPC. Sie haben sich beide geduldig um das Greenhorn von der Kunstakademie gekümmert.

Weiterhin gilt mein Dank allen, die tatkräftig am Zustandekommen dieser Arbeit geholfen haben und ohne deren fachlichen und persönlichen Beitrag alles ein wenig farbloser gewesen wäre. Dr. Elisabeth Sjöholm und ihren damaligen Team bei STFI-Packforsk, Stockholm, Schweden, danke ich für ihre Betreuung während meiner von COST E41 finanzierten short term scientific mission und für zahlreiche Anregungen zum Thema Celluloselöslichkeit. Bei Ao.Univ.Prof. Dipl.-Ing. Dr.nat.techn. Marie-Theres Hauser und Ao.Univ.Prof. Dipl.-Ing. Dr.techn. Thomas Prohaska, beide von der BOKU, bedanke ich mich herzlich für die Einführung am confocal laser scanning microscope respektive am LA-ICP-MS. Dr. Anne-Laurence Dupont vom Centre de Recherches sur la Conservation de Documents Graphiques – CNRS, Paris, Frankreich verdanke ich viele anregende Diskussionen zum Thema Nass-Trocken-Grenzflächen und interessante Proben. Bei Lic.phil. Ulrike Bürger, Leiterin Zentrum Historische Bestände, Universitätsbibliothek Bern, Schweiz, möchte ich mich für die wertvollen Kupferfraßproben und die angenehme Zusammenarbeit bedanken.

Im Rahmen des DFG-Projektes zum Thema „Restaurierung der durch Tintenfraß beschädigten Handschriften des Savigny-Nachlasses. Anwendung der Calciumphytat-Calciumhydrogencarbonat-Behandlung und partieller Stabilisierung in der Praxis" gilt mein besonderer Dank der Projektleiterin Dipl. Rest. Ulrike Hähner von der Universitätsbibliothek Marburg, Deutschland. Ein großer Dank geht auch an Dr. Rebecca Reibke, vormals an der Staatlichen Akademie der Bildenden Künste Stuttgart, Deutschland, die in die Herstellung von geeigneten Probenmaterialien viel Zeit und Mühe investiert hat. In diesem Zusammenhang möchte ich mich auch bei Dipl. Rest Enke Huhsmann, die jetzt an der Württembergischen Landesbibliothek in Stuttgart, Deutschland, arbeitet und bei Dipl. Rest Gesa Kolbe bedanken, die einen Probensatz hergestellt hat.

Die Proben zur Untersuchung des simulierten Kupferfraßes wurden zusammen mit Dipl. Rest. Katrin Schröter, die heute in der Staatlichen Graphischen Sammlung München, Deutschland arbeitet, an der Staatlichen Akademie der Bildenden Künste Stuttgart, Deutschland, hergestellt.

Allen Kollegen aus der Arbeitsgruppe der Organischen Chemie danke ich für die kollegiale Aufnahme und gute Zusammenarbeit. Was wäre schließlich eine Promotion ohne Kaffeepause - und ohne Schokolade?

Bei den umfangreichen Messungen der Papiereigenschaften haben mir Johanna Smolle, Karin Lorenz und Julian Schiehser geholfen, wofür ich mich bei ihnen herzlich bedanken möchte.

Diese Arbeit wurde teilweise von der Preservation Academy GmbH Leipzig finanziert. Ich möchte mich an dieser Stelle herzlich dafür bedanken, auch für das Bereitstellen von Testpapieren.

Zu guter letzt möchte ich mich bei meinen Eltern, meiner Oma, meinem Mann und überhaupt allen aus meinem Freundes- und Verwandtenkreis bedanken, die sich jahrelang meine Exkursionen in die Celluloseanalytik angehört haben und auch in Zukunft noch anhören müssen.

Abstract

The analysis of historical papers is a requirement to estimate their actual condition, to elucidate underlying degradation mechanisms and to design appropriate treatments. Fluorescence labelling of cellulose functionalities followed by gel permeation chromatography yields highly sensitive information about the extent of oxidation and chain scission in cellulosic materials. Additionally, only few milligrams of sample are needed which allows for sampling on selected original materials. These features make fluorescence labelling very attractive for the analysis of topics in the context of paper conservation.

Degradation processes of historic papers, their general condition in terms of average molecular weight and oxidized functionalities, the extent of oxidation and hydrolysis triggered by transition metal ions contained in pigments or inks and wet-dry interfaces formed after local wetting of paper are described on the basis of the chosen analytics on model papers and original sample material.

Various possibilities to treat some of the above described degradation processes are presented. In this work the efficiency of combined calcium phytate/ calcium hydrogen carbonate treatment to inhibit degradation caused by transition metal ions and paper deacidification on an industrial scale to prevent acid hydrolysis were investigated. The comparison between sample materials and historic paper samples is emphasized.

Typical 19th and 20th century ground wood papers are difficult to analyze due to their limited solubility in N,N-dimethylacetamide/ lithium chloride needed prior to gel permeation chromatography. Several advances to improve their solubility are described. A multivariate calibration approach based on fluorescence labelling data and near infrared spectroscopy is discussed in the present work to avoid micro-destructive sampling on historic originals.

Keywords: aging, cellulose, conservation, degradation, fluorescence

Kurzfassung

Die Untersuchung von historischen Papieren ist eine Voraussetzung für die Zustandsbeschreibung, die Aufklärung von Abbaumechanismen und die Entwicklung von passenden Behandlungen. Fluoreszenzmarkierung von funktionellen Gruppen und nachfolgende Gelpermeationschromatographie erlauben die sensitive Erfassung von Daten über das Ausmaß von Oxidation und Kettenspaltung in Cellulose. Es werden nur wenige Milligramm Material benötigt, so dass auch an Originalen Proben entnommen werden können. Diese Eigenschaften machen die Fluoreszenzmarkierung zu einem vielseitigen Werkzeug für Fragen in der Papierkonservierung.

Abbauprozesse in historischen Papieren, deren Erhaltungszustand in Bezug auf ihre durchschnittliche Molmasse und ihren Oxidationsgrad, der Anteil von Oxidation und saurer Hydrolyse, welche durch Übergangsmetallionen aus Tinten und Pigmenten verursacht werden und die Entstehung von Nass-trocken Grenzflächen nach lokaler Befeuchtung wurden mit der gewählten Analytik an Modell- und Originalpapieren untersucht.

Die Möglichkeiten zur Behandlung einiger der oben erwähnten Schäden wurden untersucht. In dieser Arbeit wurde die Wirksamkeit von einer kombinierten Calciumphytat/ Calciumhydrogencarbonat-Behandlung zur Verlangsamung von Abbauprozessen, die durch Übergangsmetallionen verursacht werden, und Massenentsäuerung von Papier zur Unterdrückung von saurer Hydrolyse überprüft. Die untersuchten Probenmaterialien werden außerdem mit Originalpapieren verglichen.

Viele Papiere des 19. und 20. Jahrhunderts sind aufgrund ihres Holzschliffgehalts mit der gewählten Analytik schwer zugänglich, da sie nur bedingt in *N,N*-dimethylacetamid/ Lithiumchlorid löslich sind. Verschiedene Ansätze zur Verbesserung der Löslichkeit werden vorgestellt. Weiterhin wird eine multivariate Kalibrierung diskutiert, die nass-chemische Daten mit Infrarotspektren korreliert und mikro-destruktive Probennahme an historischen Originalen ersetzen soll.

Schlagwörter: Abbau, Alterung, Cellulose, Fluoreszenz, Konservierung

List of own publications

1. Henniges, U., Prohaska, T., Banik, G., Potthast, A. *Cellulose* **2006**, *13(4)*, 421-428. A fluorescence labeling approach to assess the deterioration state of aged papers.
2. Henniges, U., Banik, G., Potthast, A. *Macromol. Symp.* **2006**, *232*, 129-136. Comparision of aqueous and non-aqueous treatments of cellulose to reduce copper-catalyzed oxidation processes.
3. Henniges, U., Smolle, J., Rosenau, T., Kosma, P., Potthast, A. *Lenz. Ber.* **2006**, *86*, 106-110. Mapping of aging status in historic papers by fluorescence labeling of oxidized groups and pH measurements.
4. Henniges U., Bürger U., Banik G., Rosenau T., Potthast A. *Macromol. Symp.* **2007**, *244*, 194-203. Copper corrosion: Comparison between naturally aged papers and artificially aged model papers.
5. Henniges, U., Kloser, E., Patel, A., Potthast, A.,Kosma, P., Fischer, M., Fischer, K., Rosenau, T. *Cellulose* **2007**, *5*, 497-511. Studies on DMSO-containing carbanilation mixtures: chemistry, oxidations and cellulose integrity.
6. Henniges, U., Banik, G., Reibke, R., Potthast, A. *Macromol. Symp.* **2008**, *262*, 150-161. Studies into the early degradation stages of cellulose by different iron gall ink components.
7. Henniges, U., Potthast, A. *Restaurator* **2008**, *29*, 219-234. Phytate treatment of metallo-gallate inks: Investigation of its effectiveness on models and historic samples.
8. Potthast, A., Henniges, U., Banik, G. *Cellulose* **2008**, *15*, 849-859. Ink-induced corrosion of cellulose: aging, degradation and stabilization. Part 1: model paper studies.
9. Henniges, U., Reibke, R., Banik, G., Huhsmann, E., Hähner, U., Prohaska, T., Potthast, A. *Cellulose* **2008**, *15*, 861-870. Ink-induced corrosion of cellulose: aging, degradation and stabilization. Part 2: application on historic sample material.
10. Henniges, U., Schwanninger, M., Potthast, A. Quantitative analysis of pulp hand sheets and historic papers by means of near infrared (NIR) spectroscopy and partial least square regression (PLS-R), *Carbohydrate Polymers* **2008**, doi:10.1016/j.carbpol.2008.10.028.

List of abbreviations

AGU	Anhydroglucose unit
ASTM	American Society for Testing and Materials
BKZO	TCF Magnesium Sulphite Pulp from Beech
Cadoxen	Cadmiumethylenediamine
CCOA	Carbazole-9-carboxylic Acid
CL	Cotton linters
CLSM	Confocal Laser Scanning Microscope
CNRS	Centre National de la Recherche Scientifique
COOH	Carboxyl groups
C=O	Carbonyl groups
Cuen	Cupriethylenediamine
CV	Cross Validation
d	Days
DIN	Deutsche Industrienorm
ISO	International Organization for Standardization
DMAc	N,N-Dimethylacetamide
DMAc/ LiCl	N,N-Dimethylacetamide/ Lithium Chloride
DMI	1,3-Dimethyl-2-imidazolidinone
DMSO	Dimethyl Sulfoxide
dn/dc	Refractive Index increment
DP	Degree of Polymerisation
DS	Degree of Substitution
EDTA	Ethylenediaminetetraacetic Acid
EIC	Ethyl Isocyanate
FDAM	9H-fluoren-2-yl-diazomethane
FTIR	Fourier-Transform Infrared Spectroscopy
GPC	Gel Permeation Chromatography
h	Hour
ILS	Inverse Linear Regression
IR	Infrared
LA-ICP-MS	Laser Ablation Inductively Coupled Plasma Mass Spectroscopy
MALLS	Multi Angle Laser Light Scattering
MIR	Mid Infrared Spectroscopy
M_n	Number Molecular Weight
MNSO	Methyl-(2-naphthyl) Sulfoxide
M_w	Weighted Molecular Weight
MW	Molecular Weight
MWD	Molecular Weight Distribution
M_z	z-average Molecular Weight
NIR	Near Infrared
NMMO	N-methylmorpholine-N-oxide
NMR	Nuclear Magnetic Resonance Spectroscopy
PCA	Principle Component Analysis
PDI	Polydispersity Index
Ph-NCO	Phenyl Isocyanate
PLS	Partial Least Squares
R	Correlation Coefficient
REG	Reducing End Group
RH	Relative Humidity
RI	Refractive Index
RMSEC	Root Mean Square of Calibration
RMSECV	Root Mean Square of Cross Validation
RMSEP	Root Mean Square of Prediction

SD	Standard Deviation
SEC	Size Exclusion Chromatography
STFI	Swedish Test Fibre Institute
TAPPI	Technical Association of the Pulp and Paper Industry
TBAB	Borane-*tert*-butylamine
TEMPO	2,2`,6,6`-Tetramethylpiperidine-1-oxyl
THF	Tetrahydrofuran
TS	Test Set Validation
TTC	2,3,5-Triphenyl-tetrazolium Chloride
UV	Ultraviolet
VOC	Volatile Organic Compound

List of contents

1 Introduction _____ **13**
   1.1   Paper _____ 13
       1.1.1 Cellulose _____ 14
       1.1.2 Hemicelluloses _____ 15
       1.1.3 Lignin _____ 16
       1.1.4 General aspects of paper degradation _____ 17
       1.1.5 Natural aging _____ 19
       1.1.6 Accelerated aging _____ 20
   1.2   Paper and cellulose analytics _____ 22
       1.2.1 Molecular weight distribution of cellulose _____ 22
       1.2.2 Oxidized functionalities in cellulose and their determination _____ 25
       1.2.3 Paper analytics based on micro sampling _____ 27
       1.2.4 Non-destructive paper analytics _____ 28
       1.2.5 MIR-Spectroscopy based on direct peak assignment ___ 29
       1.2.6 NIR-spectroscopy and multivariate calibration _____ 30
   1.3   Relevant topics in paper conservation _____ 35
       1.3.1 Condition rating of historical papers _____ 35
       1.3.2 Paper and transition metal ions _____ 36
       1.3.3 Bleaching in the restoration context _____ 38
       1.3.4 Wet-Dry interfaces _____ 39
       1.3.5 Mass deacidification _____ 40

2 Objectives _____ **42**

3 Materials and methods _____ **44**
   3.1   Pulps, papers and inks _____ 44
       3.1.1 Pulps _____ 44
       3.1.2 Model papers _____ 45
       3.1.3 Model inks and pigments _____ 47
       3.1.4 Historic papers _____ 48
   3.2   Treatments _____ 49
   3.3   Aging _____ 51
   3.4   Sampling _____ 53
   3.5   GPC and fluorescence labelling _____ 55
   3.6   Near Infrared Spectroscopy _____ 57
   3.7   Additional analytics _____ 58

4 Results and discussion _____ **60**
   4.1   Description of degradation in historic papers _____ 60

4.2	Influence of transition metal ions on cellulose and paper	67
4.2.1	Irongall ink corrosion	67
4.2.2	Degradation caused by copper containing pigments	74
4.2.3	Treatment options	83
4.2.4	Summary	93
4.3	Wet-dry interfaces	96
4.3.1	Wet-dry interfaces on Whatman filter paper	96
4.3.2	Wet-dry interfaces on historic papers	101
4.3.3	Summary	106
4.4	Mass deacidification	108
4.4.1	β-elimination	108
4.4.2	Sustainability of treatments	112
4.4.3	Summary	115
4.5	Improvement of cellulose solubility	116
4.5.1	Study of cellulose degradation in derivatization systems	116
4.5.2	Study of improvement of cellulose solubility	119
4.5.3	Study of the influence of EIC on pulp	123
4.5.4	Isocyanates in combination with fluorescence labelling	125
4.5.5	Summary	128
4.6	Development of a non-destructive approach	129
4.6.1	Pulp hand sheets	130
4.6.2	Historic rag papers	132
4.6.3	Summary	134
5	**Practical aspects**	**135**
6	**Summary**	**137**
7	**References**	**141**

1 Introduction

The evaluation of historical papers is a difficult task. Most paper tests have been designed to describe the properties of freshly produced papers and their suitability for further processing like printing or packaging. The main drawback of these paper industry methods for the evaluation of historical materials is obviously their huge demand on sample material. Additionally, they are often too insensitive towards early stages of degradation and subtle changes.

Nevertheless, the analysis of historical papers, mostly items of esthetical or documentary value, is highly desirable to estimate their actual condition. The prediction of future behaviour of the object relays on this estimation, too. Having data such as molecular weight, carbonyl and carboxyl group content at hand, the underlying degradation mechanisms can be revealed. These detailed data enable studies about new treatment procedures and help to monitor their efficiency.

While micro-destructive sampling on historical originals is sometimes accepted, the ideal analysis would be a non-destructive one. When no sampling is required, nor is the sample touched or altered through analysis, the investigation can be carried out directly on the object of art.

This chapter will give an introductory overview on the structure of the main paper components. Thereafter, important cellulose analytics, focussing on micro- and non-destructive methods, are presented. Relevant treatments in the context of paper conservation such as mass deacidification and degradation phenomena caused by transition metal ions will be reviewed to underline the importance of comprehensive data for further decision making.

1.1 Paper

Paper is a composite material, containing cellulose, hemicelluloses, lignin, sizing agents, mineral fillers, and other additives. It is defined as a substance composed of fibres interlaced into a compact web. In Europe, these fibres traditionally originated from annual plants that have been used for textile production. Only after excessive wear and use the old textiles were recycled for papermaking. Modern papers are mainly made out of wood fibres containing more hemicelluloses and, depending on pulping parameters, also more lignin than traditional rag papers. In general cellulose is the main constituent of paper.

Besides pure natural cellulose synthesized *e.g.* by bacteria most sources of cellulose contain important amounts of additional substances. Cellulose derived from wood is accompanied by high molecular components like hemicelluloses and lignin, and also low molecular species of organic and inorganic matter such as extractives and minerals are present [1]. In table 1, some common cellulose containing materials are reviewed for their composition. The two other polymeric constituents, hemicelluloses and lignin, play a very important role in paper making.

Table 1. Composition of some cellulose containing materials in % [2]

	Cellulose	Hemicelluloses	Lignin	Extractives
Hardwood	43 - 47	25 - 35	16 - 24	2 - 8
Softwood	40 - 44	25 - 29	25 - 31	1 - 5
Cotton	95	2	1	0.4
Flax (retted)	71	21	2	6
Flax (unretted)	63	12	3	13
Hemp	70	22	6	2

Sizing has always been necessary for paper in order to make it suitable for writing, drawing and printing purposes. In the beginning of paper production exclusively natural polymers such as starch and gelatine have been used to coat the surface of paper to impregnate it. With growing need for paper and accelerating production velocity, alum rosin sizing was used, because it could be added directly to the pulp before the sheet formation process. It turned out that this type of sizing was not very compatible with modern fillers like calcium carbonate; therefore sizing technology has been changed to alkaline sizing systems.

As fibres are the most expensive part in paper production, modern papers, *i.e.* dating from 19th century, usually contain large amounts (up to 40 %) of fillers. Other important reasons for adding fillers such as calcium carbonate, barium sulphate, gypsum, clay or talc, are the increased whiteness, opacity, smoothness and printability of filled papers. Nevertheless, in historic papers varying amounts of fillers have also to be expected [3].

1.1.1 Cellulose

Cellulose is the most abundant renewable resource. In nature, cellulose occurs in different modifications, sometimes rather pure, *e.g.* seed hair from cotton plants, sometimes in intimate contact with hemicelluloses and lignin in wood. Next to wood and annual plants for example straw, hemp or jute, cellulose is also produced by bacteria and algae in a very pure form.

Cellulose is a linear homopolymer consisting of *D*-anhydroglucopyranose units linked by 1,4-β-glucosidic bonds (figure 1). Often they are abbreviated as AGU (anhydroglucose unit). The main characteristics are three hydroxyl groups per AGU at C2, C3 and C6. There is one reducing end group (REG) at C1 and a non-reducing end group at C4 [4].

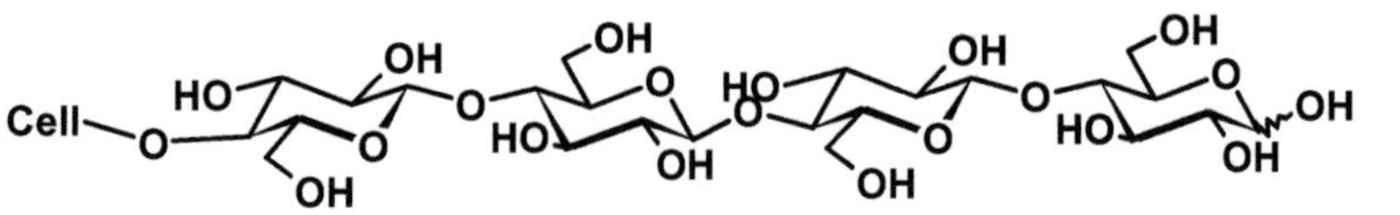

Figure 1. Molecular structure of cellulose

Being a macromolecule, cellulose contains a number of joint monomers to form chains with up to 10000 single AGUs as found in native cotton. The number of AGUs linked together is described as Degree of Polymerisation (DP). Mechanical and chemical treatments performed to transform cellulose in a suitable material for paper or textile production reduces the DP (Degree of Polymerisation) by a factor of 10 or more [4].

The polysaccharide cellulose is always polydispers, *i.e.* it has a distribution of molecules of different chain lengths. One main characteristic of cellulose is therefore the length of the molecules, either being expressed as average DP or by analyzing the molecular weight distribution (MWD). The polydispersity of the cellulose polymer is reflected by the shape of the MWD (molecular weight distribution). Due to the polydispersity of cellulose chains, statements on molecular weight (MW) are made upon statistical evaluations using different statistical moments to calculate averages [4].

Single cellulose molecules are stabilized by intra and inter molecular hydrogen bonds (figure 2). Single bonds of this type are not very strong, but the abundance of bonds and its partly higher degree of order, especially in crystalline regions, turn cellulose into a polymer that is not easily dissolved. Nevertheless for determination of MW (molecular weight) and MWD the dissolution of cellulose is required. In this context, accessibility of cellulose and how the hydrogen bonds are interrupted are key parameters [4].

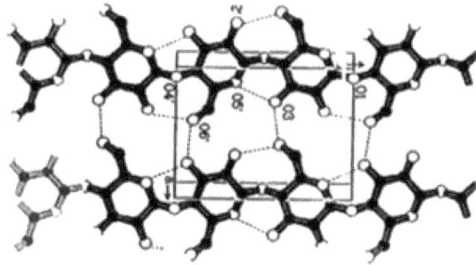

Figure 2. Supramolecular structure of cellulose [5]

1.1.2 Hemicelluloses

Hemicelluloses are polymers with a lower MW (between DP 50 and 200) than cellulose. They occur in a large variety of structural types, divided into four general groups [6]:
- Xyloglycans (Xylans), the most abundant hemicellulose type
- Mannoglycans (Mannans)
- Xyloglucans, representing the major constituent of the primary cell wall of all higher plants
- Mixed linkage β-glucans, also known as cereal β-glucans

Xylans are subdivided into homoxylans with differently linked xylose units, glucuronoxylans with linked 4-O-methylglucoronic acids (figure 3), (arabino)glucuronoxylans, arabinoxylans and (glucurono)arabinoxylans The last one, (D-glucurono)-L-arabino-D-xylans are reported to be the dominant hemicelluloses in the lignified tissues of some annual plants [7]. In general, xylans are typically found in hardwoods.

Mannans are built out of mannose and glucose units. They are subdivided into galactomannans that differ in their degree of branching, glucomannans and galactoglucomannans [7]. Softwoods mainly contain a high portion of mannans.

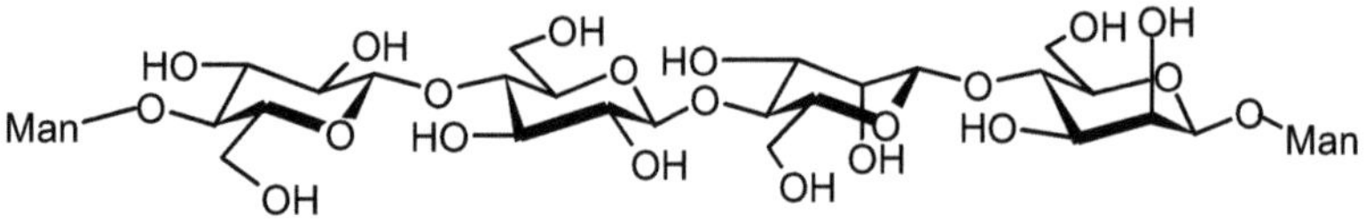

Figure 3. Structure of one hemicellulose, D-gluco-D-mannan.

While cellulose molecules form regular structures and are organized in elementary fibrils, hemicelluloses help to partly aggregate these arranged cellulose molecules into microfibrils. These microfibrils form macrofibrils having again hemicelluloses and additionally lignin linking them together. Thus, a close network between hemicelluloses and other wood components is formed, partly stabilised by chemical bonds, partly by hydrogen bonds [8].

In paper making hemicelluloses are appreciated because they add to yield and contribute to fibre-fibre bonds. Additionally, they provide flexibility to the cellulose fibres. This feature was explained by internal stress redistribution and increased swelling ability that hinders hornification of fibres [9, 10]. In paper conservation the role of hemicellulose on long term stability and its behaviour towards conservation treatments is not very much discussed until now. Regarding its stability within the paper network, hemicelluloses will definitively introduce acid components into the paper, thus accelerating aging and deterioration processes. Under acidic conditions, hemicelluloses hydrolyze at a much greater rate than cellulose [11]. Additionally, they are easier accessible than cellulose is. On the other hand, carboxyl groups at C6 position decrease the rate of hydrolysis under acid conditions, while the opposite is true for an aldehyde group at the same position [12]. An evaluation of the role of hemicelluloses in the context of paper conservation is still lacking.

1.1.3 Lignin

Lignin, the strengthening agent in wood, is characteristic for higher plants. It typically occurs in vascular tissues. In contrast to cellulose and hemicelluloses, lignin is rather hydrophobic; therefore it enables transport of liquids within the living plant. Biosynthesis of lignin starts from a glucose unit, but unlike cellulose and hemicelluloses, the resulting molecule is a three dimensional network of phenyl propane units that are connected to each other with different bonds. There are three main building units, *p*-coumaryl alcohol, coniferyl alcohol and sinapyl alcohol (figure 4). Their ratios and occurrence differ; softwood lignin mainly contains coniferyl alcohol, hardwood lignin is built by both, coniferyl and sinapyl alcohols, and the lignin of annual plants like grass is characterized by the presence of *p*-coumaryl alcohol [1].

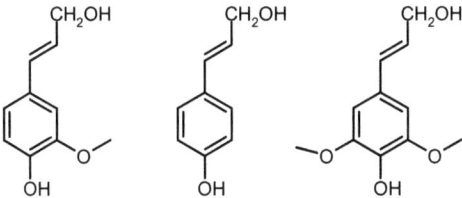

Figure 4. Lignin precursors. Left: trans-coniferyl alcohol. Middle: trans-p-coumaryl alcohol. Right: trans-sinapyl alcohol.

Being the most complex molecule in the context of cell wall polymers, the elucidation of its structure and main bonds has been pursuit for a long time using different approaches. It is generally accepted today, that lignin is not simply deposited between cell wall polysaccharides, but directly linked to at least a part of them. These intimate associations are called lignin-polysaccharide complexes [1].

Except for ground wood and mechanical pulp where lignin is not removed, pulp processing aims at the removal and modification of lignin [1]. For paper conservation lignin content is usually regarded to be a disadvantage. It yields a multitude of degradation products upon oxidative and hydrolytic degradation [13]. Additionally, the whole macromolecule is sensitive towards UV-irradiation with absorption maxima around 280 nm [1]. While some authors claim that lignin does not reduce the mechanical strength of paper [14], it was shown that photo oxidation can be triggered by lignin [15].

Despite of all these negative and unresolved aspects of the suitability of lignin in the context of paper permanence, it was argumented that paper production has been improved considerably in the last decades. The use of chemicals has become more controlled, and alkaline fillers have gained more importance. Therefore, the maintenance of at least low amounts of lignin in paper has been rehabilitated to some degree with respect to paper permanence [16]. Some more recent opinions tend to charge historic pulping and paper production conditions for bad permanence of papers from the 19th and 20th century rather than lignin content [17].

1.1.4 General aspects of paper degradation

When dealing with cellulose in the context of modern and historic paper, degradation is an important issue. The degradation process is visually perceived by yellowing, *i.e.* change in colour, and embrittlement, thus a change in mechanical properties. Depending on the purpose of the paper document, none, either of them or both degradation phenomena may be acceptable. The process of yellowing and embrittlement is also defined as aging in the context of paper conservation.

Main degradation pathways for cellulose are hydrolytic chain cleavage caused by acids or enzymes, degradation caused by alkali in aqueous medium, oxidative, mechanical, thermal and radiation degradation of cellulose. For historic papers, acid-catalyzed hydrolysis of cellulose is the most important degradation reaction (figure 5). Nevertheless, oxidative processes may also play an important role when the pulp contains traces of metal ions (figure 6). Both are degradation processes inherent to a natural polymer, but can be considerably accelerated by various factors.

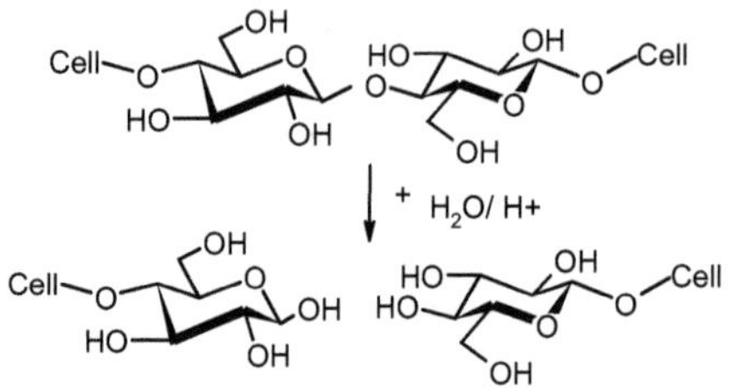

Figure 5. Acid hydrolysis of cellulose

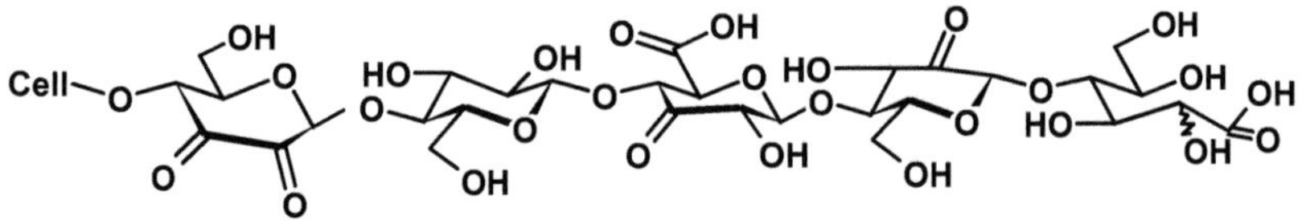

Figure 6. Cellulose molecule with different oxidized functionalities

An important decline in paper producing quality started when the need to compensate for the rising demand of raw material developed. More and more chemicals such as potassium aluminium sulphate, and bleaching agents based on chlorine compounds have been used. Materials of inferior quality where thereby made available as a source for paper production [18] even before the introduction of low quality ground wood pulp. Paper from the mid 19th till the late 20th century has mainly been produced according to acid sulphite cooking processes [19]. These production parameters mainly lead to acid hydrolysis of carbohydrates and the formation of new reducing end groups [12]. Alum rosin sizing that was added to the pulp slurry provides another permanent source of acidity. Altogether these production parameters will lead to increased brittleness of the paper. Usually those papers from the last two centuries contain significant amounts of lignin as well, which can be oxidized more easily. Besides these inherent parameters of paper degradation, storing conditions will also influence the aging behaviour. All these degradation reactions are reflected by changes in molecular weight, molecular weight distribution and cellulose functionalities. Using appropriate methods to detect these modifications yields information on the type of degradation [4].

On a molecular level aging can be monitored by measuring the extent of cellulose chain scission and the formation of oxidized functionalities [20]. While oxidized functionalities are mainly related to optical properties, the length of cellulose chains translates into mechanical strength [21], even though strength properties are additionally strongly related to structures on a fibrillar level. In general, the amount of hydroxyl groups in cellulose is reduced while carbonyl and carboxyl groups increase. The shortening of average cellulose chain length is largely correlating to this increase of oxidized functionalities as there are strong interrelations between these two parameters. Nevertheless, exceptions from this rule do exist; oxidation does not necessarily lead to chain scission and during purely hydrolytic degradation, no keto and carboxyl groups are formed.

To understand the aging mechanism on a macromolecular level *i.e.* regarding the properties of the paper sheet, a concept has been put forward that describes the failure in mechanical properties by either of the following mechanisms. The first mechanism describes a uniformly distributed bond area that holds the inflicted load while fibres itself will burst. The other mechanism implies that strong fibres are only loosely bond and upon stress not the fibres will burst but the bonds fail. This concept is based on the recognition that paper strength is directly linked to fibre strength [11]. When analyzing the degradation of paper it has to be kept in mind, that paper is a composite material, consisting of very stable materials under moderate conditions, like its main constituent cellulose, while other components rather increase the rate of aging, at least under poor storage conditions.

1.1.5 Natural aging

When the inherent degradation pathways of polymers as described in 1.1.4 moderately occur in the course of decades and centuries the process is called natural aging in the context of paper conservation. Aging processes of natural polymers comprise the influence of irradiation from light, mechanical wear due to use, mechanical damage inflicted by improper handling, biodegradation caused by micro organisms and chemical changes originating from air pollution, changing climatic conditions, inappropriate raw materials or detrimental additives [22]. A classification of aging procedures can be achieved by dividing them into exogenous and endogenous processes [23]. Most of the processes listed above are exogenous influences, but especially unsuitable raw materials or sensitive additives are endogenous ones (figure 7).

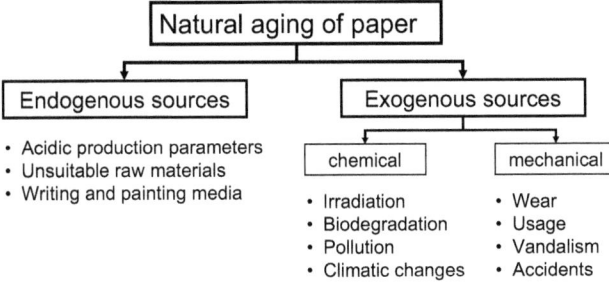

Figure 7. Classification of aging procedures

During aging, more than one chemical process has to be expected. While some compounds in paper may be bleached upon light exposure, others may turn yellow. It is generally expected that during aging a loss in MW takes place, but due to an increasing amount of carbonyl groups cross linking effects may occur. Some of these changes will even influence the chosen analytics and mislead the interpretation of results [22]. Following phenomenon might serve as an example for possible misinterpretations: high carbonyl group content in cellulose exerts a negative influence on the determination of viscosity by means of strongly alkaline cellulose solvents and leads to an underestimation of DP.

Besides all the chemical degradation processes outlined above, mechanical wear and damage due to improper handling and vandalism are man-made degradation processes not depending on chemical parameters. They can only be controlled by restricted access and education of users.

The power of solar irradiation on paper has been known for a long time, since sunlight has been used as a bleaching agent, and in some traditional producing sites it still is. The natural light source sun emits a full spectral light spectrum ranging from the ultraviolet (UV) region below 380 nm up to the infrared red (IR) region with more than 800 nm. To the human eye, only a small portion between 380 and 750 nm is visible. Paper made from high yield pulp, a mixture of cellulose, hemicelluloses and lignin yellow on exposure to light. Wavelengths below 306 nm are responsible for yellowing and longer wavelengths lead to bleaching [24]. Light induced degradation has been separated into two reaction pathways. The first is photolysis due to direct energy uptake that leads to chain scission. The second is a process that was triggered by light energy uptake resulting in a later degradation caused by radicals that were formed during energy uptake [22]. Especially carbonyl groups react in a very sensitive way towards UV energy absorption.

Other sources of irradiation are treatments using high energy to modify cellulose or paper. In the context of treating objects suffering form biological infestations, γ-radiation and its influence on paper and further paper deterioration has been discussed vividly. Nevertheless, such radiation treatments are not part of any naturally occurring aging processes, and will therefore not be discussed in this context.

Biodegradation goes hand in hand with improper storage conditions (too humid, too dirty, too warm, etc.) that also give rise to climatic changes that are detrimental to cellulose. Additionally, most other components of paper like hemicelluloses, lignin, and pigments, or starch and gelatine as binding or sizing agents are prone to biodegradation and serve as nutrients for micro organisms. In fact, most micro organisms found on paper will not directly destroy the cellulose, but rather digest other components found on or in paper. Their growth is strongly dependent on ambient conditions like temperature, relative humidity, availability of nutrients and pH. During biodegradation coloured pigments are formed that visually distort the appearance of an object. More dangerous in terms of conservation are the production of organic acids, enzymes and other degradation products during microbiotic activity [25].

Air pollution is a rather modern problem mainly related to noxious gases like sulphur dioxide, nitrogen dioxide and ozone. These gases contained in polluted air will be adsorbed on paper where some of them can form acids due to the water content of paper. Consequently, polluted air will trigger acid degradation and lead to more brittle paper [26]. In this context it has already been discussed that the major action against acid degradation of paper, so called deacidification by treatment with alkaline substances to neutralize acids in paper and deposit an alkaline reserve, might adversely lead to an increased uptake of pollution agents [27].

Mechanical damage arises from a force (stress) that causes movement (strain) that exceeds the natural extensibility of the material. In everyday life the words stress and strain are used interchangeably. In the technical context, however, there are precise definitions of these words [28]. The most immediate result of wildly fluctuating relative humidity or temperature is mechanical damage: flaking, cracking and warping. Chemical effects are much slower to appear and even biological attack is not so immediately apparent as mechanical damage [28]. Nevertheless, it was shown in a model study, that fluctuating relative humidity lead to a significant decrease in the degree of polymerisation caused by hydrolysis after 12 month [29].

1.1.6 Accelerated aging

According to Feller, accelerated aging tests are performed for three main reasons [22]. The first aim is to rank various materials with respect to their chemical stability and physical durability in a relatively short time. Secondly estimation upon long term behaviour of a material is intended. The third reason is the possibility to study degradation mechanisms in the laboratory under controlled conditions.

In a study on the rate of paper degradation, the following definition for accelerated aging methods was put forward [30]:

"For these artificial or accelerated aging methods, a material is exposed in a climate-chamber to extreme conditions in terms of temperature and humidity for a certain period of time, during which the changes occurring in the material are measured. Subsequently, from the data obtained in this way, the rate at which the material in question will deteriorate under normal, natural conditions of storage is deduced."

Without any doubt, the main obstacle in all approaches based on accelerated aging is lacking knowledge about what exactly "natural" aging is. The idea of accelerating aging processes implies some kind of uniformity concerning the degradation path and its velocity.

Additionally, if this process of natural aging obeys some standard rules the question remains if they are reproducible only be increasing the temperature and without any undesired side reactions gaining too much importance and distorting the picture [30].

In early publications mainly dry accelerated aging conditions have been used and translated directly into year equivalents [31]. In the 1960's the concept of relative humidity was introduced and critically discussed [11]. Nowadays, scientists working in the field of paper conservation have agreed upon the need to include humidity into accelerated aging set-ups. In the course of the years, many procedures for accelerated aging have been used (table 2), but only some of them have been standardized (table 3).

Table 2. Overview of different accelerated aging protocols (adopted from [30])

Method	General description and reference
80°C, 65% RH, 6-24 d	Constant temperature and relative humidity[32]
90°C, 35/80% RH (change in RH every 3 hours)	Dynamic aging (changes in RH) at elevated temperatures[33]
80°C, 10-81 d, paper samples in sealed glass tubes	Aging in closed vessels, with constant or cycling temperature[34]
80°C, 60% RH, 7-28 d; gamma radiation	Controlled exposure to irradiation[35]
80°C, 95% RH, 1-32 d; 22°C, 65% RH, SO_2 and NO_x, 1-28 d	Controlled air pollution[36]

Table 3. Overview of aging protocols to accelerate changes in paper according to different standard procedures

Method	General description	Reference
80°C, 65% RH	Constant temperature and relative humidity	DIN ISO 5630-3 and ISO 5630-3: 1996
90°C, 50% RH	Constant temperature and relative humidity	TAPPI T 544 sp 97
105°C	Dry heat	TAPPI T 453 sp 97, DIN ISO 5630-1 and ISO 5630-1:1991

Additionally, kinetics have been calculated, e.g. kinetics of cellulose degradation are given by 1/ DP expressing the number of bonds broken (Ekenstam equation). Of course, a kinetic approach relays on accelerated aging, and raising the temperature will not necessarily speed up all reaction steps to the same extent. This kinetic approach to compare paper degradation processes is not generally agreed upon. Paper, a complex mixture of different components, generally deteriorates only slowly. Especially the initial degradation steps are very difficult to capture, resulting in difficulties to extrapolate results gained at elevated temperatures to ambient conditions [22]. Especially Porck discussed this topic in detail, mainly criticizing that Arrhenius plots need to many prerequisites to be met, and that only few projects exist that really tried to verify the relevance of accelerated aging [30]. General knowledge is lacking to describe natural aging processes in a sensitive and characteristic way and to prove if this process can be simulated according to the description found for naturally aged papers.

According to Porck, accelerated aging is only relevant for direct comparison of certain paper conservation treatments, fundamental research into paper degradation mechanisms and the simulation of specific types of paper damage [30].

1.2 Paper and cellulose analytics

Usually the amount of sample needed for pulp and paper testing for industrial purpose is of no big interest. Common pulp and paper testing methods require at least 1 g of paper or, when it comes to mechanical paper testing, even several sheets of paper are needed [37]. As long as pulp and paper is produced on a ton scale these amounts do not matter. Nevertheless wet chemical analysis is time and therefore money consuming. This fact makes fast qualitative and quantitative testing attractive for the pulp and paper industry as well.

In the context of testing procedures related to conservation sample material is difficult to obtain. When it comes to destructive methods, one possibility is to investigate conservation relevant topics on freshly produced mock ups. Otherwise scarce historic material has to be used. Nevertheless, large amounts of sample material are not easily acquired as it is hardly ever found that a sufficient quantity of an appropriate material can be provided from an archive or library. Preparing mock ups is expensive and does not necessarily reflect naturally aged material.

Therefore, Porck and Teygeler [38] stressed in an overview of conservation developments that the development of micro-destructive analytical tools, or non-destructive analysis is strongly recommended in science dedicated to conservation, especially in order to monitor natural aging. In the following section, several methods for chemical cellulose analysis will be shortly reviewed, stressing micro or non-destructive approaches.

1.2.1 Molecular weight distribution of cellulose

From a historical point of view, degree of polymerisation (DP) is the first available parameter for cellulose characterization. Cellulose was dissolved and the viscosity of this solution was determined. The obtained value was than transformed mathematically into the degree of polymerisation. Common solvents were cuprammonium hydroxide solution (CUAM), cupriethylene diamine (CUEN), cadmiumethylenediamine (Cadoxen) and cold, concentrated phosphoric acid. Of course, all these methods did not result in a distribution of molecular weight, but one single average value that roughly described the DP of cellulose. It was well known, though, that all these methods would cause degradation of cellulose due to their strong alkalinity or acidity [39].

Dissolution of cellulose by derivatization

Obtaining a molecular weight distribution of cellulose was first achieved in the 1980's with gel permeation chromatography (GPC) as a separation system. For this technique the cellulose had to be derivatized to enable dissolution. The derivatives were subsequently separated chromatographically in an organic solvent to obtain their distributions [40, 41]. Carbanilation of cellulose is a very common way for cellulose derivatization and dissolution in organic solvents, mostly tetrahydrofuran (THF) [42] for the determination of analytical parameters of celluloses by GPC. During carbanilation all hydroxyl groups of cellulose should be modified, a process that disturbs inter and intramolecular hydrogen bond formation and facilitates therefore cellulose dissolution (figure 8).

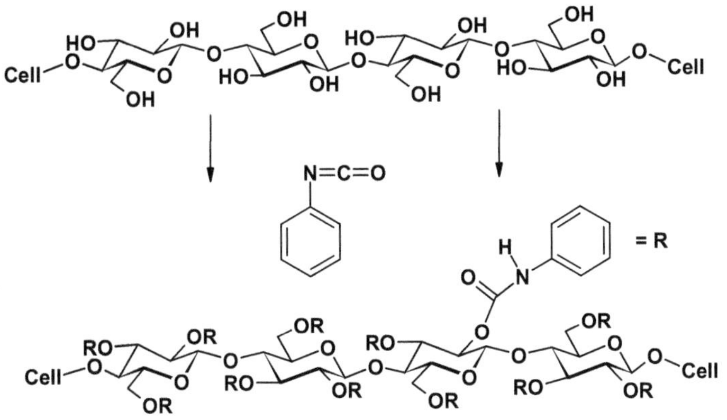

Figure 8. Derivatization of cellulose by phenyl isocyanate (= carbanilation)

Most protocols recommend dimethyl sulfoxide (DMSO) as the derivatization solvent, some suggest pyridine, or mixtures of both solvents. It is suspected that DMSO, even though being the more efficient reaction medium will lead to more degradation of cellulose as well [43, 44]. The degradation in DMSO would lead to carbonyl structures that subsequently cause chain cleavage according to β-elimination mechanisms [45]. Carbanilates are quite stable and can be stored over extended periods of time. The mobile phase used in GPC of carbanilates, tetrahydrofuran (THF), is more common and requires less specified equipment as compared to N,N-dimethylacetamide/ lithium chloride (DMAc/LiCl). Nevertheless, for chromatography purposes THF has to be used without stabilizers. As it is an ether species it may form peroxides. The increase of the polymer mass by carbanilation in combination with the large refractive index increment (dn/dc) accounts for increased sensitivity upon light-scattering detection and refractive index detection, respectively. In addition, the introduction of aromatic moieties upon derivatization using phenyl isocyanate allows for UV-detection. One obstacle to a broad and general application of carbanilation in cellulose analytics is a certain discrimination effect: low-molecular weight parts are lost during work-up and purification by precipitation and re-dissolution steps.

Direct dissolution of cellulose

During the 1980's, the solvent system N,N-dimethylacetamide/ lithium chloride (DMAc/ LiCl) was introduced into cellulose characterization by gel permeation chromatography and soon became a very popular and versatile system. Cellulose is directly solved without any derivatization step. A clear, viscose solution is formed which is highly suitable for the GPC system [46]. Another advantage is the relatively high stability of the cellulose in the solvent system; no significant degradation has been found when applied properly [47]. Also oxidized cellulose is stable. Heating or refluxing insoluble samples in DMAc/LiCl does not lead to improved solubility but causes progressing degradation of the cellulosic material [48]. This type of analysis, based on cellulose dissolved in DMAc/ LiCl and subsequent separation in a GPC has been studied thoroughly during the last years [49-51]. Most dissolving pulps, bleached paper pulps and historic rag papers are soluble in the DMAc/ LiCl system. However, when it comes to high molecular weight paper pulps and papers containing larger amounts of hemicellulose or lignin, the solubility in DMAc/LiCl is a limiting factor. A procedure to overcome this problem is based on partial derivatization with ethyl isocyanate (EIC), thus combining direct dissolution with derivatization. The derivatization with this UV innocent reagent enhances the solubility in a way that GPC of softwood Kraft pulps becomes possible, even in the presence of relatively large amounts of lignin [52].

Methods for molecular weight determination

Methods for molecular weight determination are divided into three different categories. There are absolute methods, equivalent methods and relative methods. Relative methods, like viscosimetry, depend on interactions between the chemical and physical structure of the dissolved macromolecule and additionally interactions between these properties and the used solvent system. Therefore, calibration is needed to take these relations into account. When information is available on characteristic properties of a molecule, like end groups in a polymer chain, which are equivalent to the molecular weight, than the determination of the end groups may be used for certain polymers for the calculation of number molecular weight (M_n) [53]. Most preferred methods are absolute methods that do not rely on any assumptions about chemical and physical structure properties of a molecule, but have a direct relation between measured value and desired information, *i.e.* molecular weight. All light scattering techniques, ultra centrifugation and methods depending on colligative properties[1] can be used for absolute determination of molecular weight [53]. When combining gel permeation chromatography and light scattering the molecular weight distribution plus detailed information on the structure of the molecule in solution are obtained.

Gel permeation chromatography (GPC)

The name gel permeation chromatography (GPC) originates from the soft gels that have been used for water soluble polymer characterization. Today, the name size exclusion chromatography (SEC) is interchangeably used. It refers to the application of organic gels or inorganic gel matrices. In general, all soluble polymers can be analyzed. In GPC, porous gels are used for the separation of the dissolved molecules in the columns. Molecules are separated according to their hydrodynamic volume. Therefore, molecular weight is not the basic separation criteria, but the volume that is occupied by a polymer in solution. As a consequence, an equal molecular weight of different molecules does not mean, that these molecules will elute at the same time, while equal elution time does not mean equal molecular weight. Molecules with a small hydrodynamic volume are capable of entering a huge number of pores, resulting in a long persistence in the column. These molecules will elute last. The larger the hydrodynamic volume gets, the less pores are available for penetration and the faster these molecules will pass the column. Large molecules will elute first.

A prerequisite for molecule separation in GPC is obviously dissolution. The mobile phase and the solvent that dissolves the polymer are ideally the same and do not exhibit any detrimental influence or interaction on the column packing. Most commonly cross linked polystyrene is used as column packing [54].

Light scattering for molecular weight determination

Light scattering is one of the few absolute methods for determination of molecular weight and structure. The theory of light scattering is a complex model based on the Tyndall effect that describes light scattering at colloidal dissolved molecules. A comprehensive overview of light scattering theory is given by Wyatt [55].

The average molecular weight can be calculated according to the Zimm equation:

$$K^*_c/R(\theta,c) = (1/M_w \cdot P(\theta)) + 2A_2c$$

- $R(\theta,c)$ is the excess Rayleigh ratio of the solution as a function of scattering angle θ and concentration c. It is directly proportional to the intensity of the scattered light in excess of the light scattered by the pure solvent.
- c is the solute concentration that is determined by refractive index detector.
- M_w is the weight-averaged solute molar mass.

---

[1] Colligative: The amount, not the type of dissolved polymer influences the properties of its solvent, e.g. freeze point, boiling point, osmotic pressure and vapour pressure.

- A_2 is the second virial coefficient in the virial expansion of the osmotic pressure.
- K^* is the optical constant calculated according to $4\pi^2 n_0^2 (dn/dc)^2/(N_A \lambda_0^4)$. In this equation n_0 is the refractive index of the solvent, N_A is Avogadro's number, λ_0 is the wavelength of the incident light, and dn/dc is the refractive index increment. Note that dn/dc is squared in the equation; therefore it has to be determined carefully.
- $P(\theta)$ describes the angular dependence of the scattered light, and can be related to the rms radius.

Experimentally, the Zimm equation is solved by plotting a variety of scattering angles, θ, at a constant concentration. In figure 10 an example of this procedure is given. On the x-axis all available scattering angles, in this example 17 different ones, are plotted as $\sin^2(\theta/2)$ and extrapolated to the y-axis. According to Zimm equation the inverse of M_w is obtained when solving K^*c/R (theta). Graphically this solution is expressed by the intercept of the extrapolated scattering angles with the y-axis (figure 9).

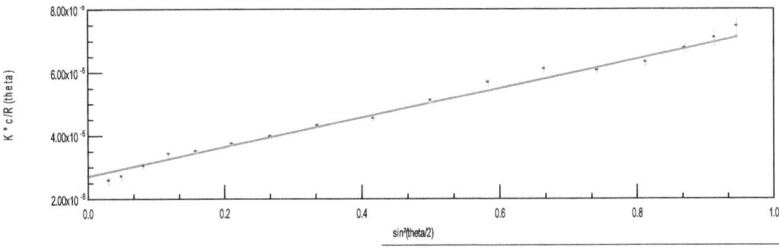

Figure 9. Experimental solution of the Zimm equation via plotting all available scattering angles on the x-axis and extrapolating them to y-axis.

1.2.2 Oxidized functionalities in cellulose and their determination

Next to the molecular weight, other main parameters of a polymer are its substituents and functional groups. Native cellulose, as for example produced by bacteria, will only have one reducing end group per chain. All the other functional groups are hydroxyl groups, three per anhydroglucose units (AGU). Due to natural aging, irradiation and chemical processing, e.g. pulping and bleaching, an increasing part of these hydroxyl groups will be transformed into oxidized functionalities, i.e. aldehydes, ketones and carboxylic groups (figure 10). The amount of these oxidized functionalities is small (µmol/ g), though, and therefore, it is difficult to perform carbonyl and carboxyl group determination [56].

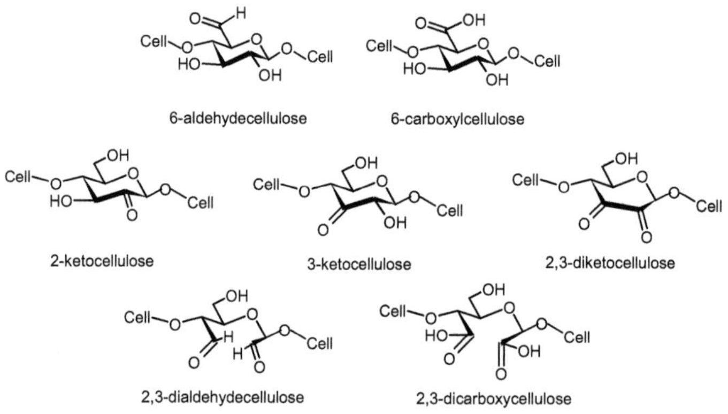

Figure 10. Different oxidized cellulose structures (according to [4])

Carbonyl groups in cellulose

Next to the naturally occurring reducing end group (REG), further carbonyl groups can be formed most easily at C6 (6-aldehyde structures). Keto groups can occur at C2 or C3 or both (2,3-diketo structures). Further aldehyde functional groups may be formed at C2 and C3 by periodate oxidation upon ring opening (2,3-dialdehyde structures). All types of carbonyl groups induce β-elimination that will lead to chain scission of the cellulose polymer in alkaline conditions. Carbonyl functions in cellulose are present to a large extent as their hydrates and/ or hemiacetals/ hemiketals next to their double-bonded form [57]. Thus, they temporarily do not exhibit their double-bond characteristics, and hindering their detection by spectroscopic methods like FTIR [56]. Additionally, their concentration is normally very low. Conventional spectroscopic methods such as IR, UV and NMR are not sufficiently sensitive to yield reliable quantitative results [58].

There are several methods to determine the carbonyl group content in cellulose, mainly relying on titration or photometric detection. Most of them are considered to be rather insensitive, some of them even only semi quantitative. So far, no direct method is available that may differentiate between keto and aldehyde functionalities [56]. For pulp and paper industry a valuable method is the copper number, defined as the number of grams of metallic copper resulting from the reduction of $CuSO_4$ by 100g of pulp or paper fibres (TAPPI T 430 om-94). It is reported that it mainly reacts with aldehyde functionalities, though not exclusively [58], while the whole reaction mechanism is still not fully understood and remains ill-defined [56]. Oximation by reaction with hydroxyl amine followed by different detection methods using hydroxylamine is another frequently used approach. It is considered to be the most appropriate method among the classical methods with its sensitivity depending on the detection method. Further methods are sodium borohydride reduction of carbonyl functionalities and determination of released hydrogen gas, the formation of cyanohydrins by reaction of carbonyl groups with cyanide and subsequent titration with silver nitrate or determination of nitrogen content and an approach based on 2,3,5-triphenyl-tetrazolium chloride (TTC method). The TTC method determines photometrically the amount of the red dye that is formed by the reaction of TTC with reducing end groups. The cyanide-method is problematic due to its toxicity, and often leads to an overestimation of oxidized functionalities due to adsorption phenomena. TTC leads to cellulose degradation and the sodium borohydride reduction is considered to be rather insensitive.

A novel method to accurately determine the carbonyl content in cellulosic materials by selective fluorescence labelling with carbazole-9-carboxylic acid [2-(2-aminooxyethoxy)ethoxy]amide (CCOA) was developed recently. The procedure can readily be incorporated into a gel permeation chromatography (GPC) system with refractive index and multiple-angle laser light scattering detection. A procedure for combining the labelling approach, using aqueous buffer pH 4.0, and direct dissolution of the labelled pulp in DMAc/LiCl 9 % (w/v) for determination of carbonyls in pulps was optimized with regard to reaction conditions, presence of catalysts, reproducibility, and completeness of conversion. This method is limited by the necessity to dissolve pulp or paper prior to analysis. The CCOA method has the advantage of high sensitivity and high selectivity while only very small sample amounts are needed. The method helps to overcome the drawback that only sum parameters are accessible, and its broad applicability has been demonstrated [59-61].

Carboxyl groups in cellulose

In contrast to carbonyl structures, carboxyl groups are not naturally contained in cellulose. They are all formed due to oxidative stress. When the reducing end group is oxidized, gluconic acid is formed. The formation of this acid hinders further alkaline peeling starting from the reducing end group and is therefore considered to be beneficial to the cellulose. When C6 of the AGU is further oxidized, glucuronic acid is formed. This modification can be achieved by TEMPO oxidation or nitrogen oxide treatments. When acids are formed at C2 or C3, the ring structure will be destroyed. A major source of carboxyl groups originates from residual hemicelluloses. Especially in hardwood xylans considerable amounts of *4-O-methylglucuronic acids* are contained.

There are several methods for the determination of carboxyl groups generally based on two concepts. The carboxyl group content can be determined by alkalimetric titration after converting the carboxylate functions into free COOH groups. Another option is the binding of suitable cations, assessing the amount bound or the decrease in concentration in the solution [4]. The only method standardized by TAPPI works via consumption of $NaHCO_3$ [58]. In this procedure, the pulp is extracted with dilute hydrochloric acid, washed, reacted with sodium hydrogen carbonate-sodium chloride solution, and filtered. The filtrate is titrated with 0.01 N hydrochloric acid to methyl red end point (TAPPI T237 om-93). The methylene blue method was abandoned by TAPPI as it was considered to be too time consuming. Nevertheless in a study on available methods for carboxyl group determination, it was concluded that it is a very suitable method yielding high repeatability [62]. As both approaches depend on the accessibility of suspended fibres, their completeness was discussed. It was found that about 85 to 100 % of the total carboxyl group content is available under these conditions [4]. A more recent approach is based on photo acoustic rapid-scan Fourier transform infrared spectroscopy in the mid-infrared region [63].

Like for carbonyl groups, a selective fluorescence labelling of carboxyl groups and subsequent GPC coupled with fluorescence detection was developed. A novel fluorescence label, 9H-fluoren-2-yl-diazomethane (FDAM) was synthesized. Applying the FDAM method, for the first time, carboxyl profiles of cellulosics, *i.e.*, the carboxyl content relative to the molecular weight distribution, were obtained [64].

1.2.3 Paper analytics based on micro sampling

Scientists working in the field of paper conservation have made big efforts and advances in reducing the amount of sample material needed in order to make investigations on original material, work of arts and documents feasible. Some of the most relevant methods are presented in this section.

Determination of pH

As the pH of paper is considered to be a very important parameter to decide upon the need and success of deacidification treatments, several techniques for the micro determination of pH have been suggested [65, 66]. Unfortunately, pH does not yield any information about strength properties or undergone oxidation; it allows only a rough estimate on the future perspectives of stability. Another drawback is the often observed effect, that in the acid range pH measurements are of more accurate than in the alkaline range [66]. Additionally, pH cannot be correlated to the amount of alkaline reserve, *i.e.* it is very doubtful if pH will really tell about the success of a deacidification treatment, because buffering effects of other paper constituents are expected to lead to a levelling off effect [65]. Surface pH is more difficult to interpret than the extract approach; it probably only reflects parts of the acidity and alkalinity contained in paper. According to standard procedures the measurements do not take long time[2]. This implies that only components of paper that dissolve rapidly will contribute to the reading. Even though pH of paper does not give a very detailed insight into the amount of degradation processes going on and it is not a very sensitive figure, it helps to deduce to the kind of degradation occurring as for example acid degradation is not likely in alkaline paper.

Size exclusion electro chromatography (SEEC)

Another suggestion was the use of micro SEEC on single paper fibres to assess the deterioration state of works of historic or esthetical value on paper [67]. In SEEC, an analytical technique belonging to the group of capillary electrophoresis, the mobile phase is driven by means of electro osmosis, which develops upon the application of a high electric field across the capillary column. Typical column dimensions of SEEC are inner diameters of 50 - 100 μm and lengths of 0.25 - 0.5 m. This allows for the considerable reduction of sample and mobile-phase consumption compared to conventional GPC. The peculiarities of electro osmotic flow (EOF) allow for an improvement of separation efficiency as well. Moreover, the use of smaller particles to generate higher separation efficiencies is more easily realized in SEEC than in pressure driven chromatography. On the other hand, the mass selectivity in SEEC may be lower than in GPC. Using SEEC on single paper fibres the authors themselves have noted one big drawback, namely that different fibres of a single object would give different results even though the method as such proved to be reproducible on one and the same fibre.

1.2.4 Non-destructive paper analytics

Chemiluminescence

The application of chemiluminescence for non destructive analysis was suggested to assess the various characteristics of paper [68]. Chemiluminescence has been defined as light emission originating from the relaxation of electrons which populate excited states in an elementary process of chemical reaction. In general, chemiluminometry is considered to be a complementary method to study the role of oxidized functionalities and radicals during oxidative polymer degradation. Especially the formation of radicals can easily be followed by chemiluminometry. It was suggested on the base of chemiluminometry that swelling of cellulose due to water absorption will lead to bond scission with subsequent increased oxidation [69].

Recent studies focussed on the applicability for cellulose analysis. Several correlations have been found regarding oxidation and non-oxidation reactions. A correlation between the decrease of DP and an integrated chemiluminescence signal at different temperatures under nitrogen atmosphere and isothermal heating was shown. Consequently, also non-oxidation phenomena can be studied using chemiluminometry [70]. However, most investigations have been dedicated to oxidation. There are correlations between peroxide and aldehyde group

---

[2] For example TAPPI T 529 om-04 requires at least 2 minutes and claims that no longer than 30 minutes are necessary to reach equilibrium.

content [71], between peroxide content and the rate of chain scission [72] and between peak height of chemiluminescence after heating in nitrogen and carbonyl group content [71].

However, chemiluminescence is a complex process and the emitted light is not selective for one defined species. Chemiluminescence depends upon polymer characteristics, such as previous oxidation, temperature, water, and additives to the studied polymer, *e.g.* paper additives when analyzing cellulose, which may mislead interpretation [73].

Solid phase micro extraction coupled with gas chromatography and mass spectrometry (SPME-GC-MS)

Another method introduced in paper conservation studies has been solid phase micro extraction coupled with gas chromatography and mass spectrometry (SPME-GC-MS). The highly sensitive SPME fibres have been used to trap volatile degradation products of paper and after desorption GC-MS has been used to analyse these substances. Attempts have been made to define marker substances that yield information on the deterioration state of the investigated paper. There is agreement by several researchers that furfural will indicate acid hydrolysis while acetic acid is an indicator for oxidative degradation of cellulose. There is even indication that calibration of these indicator substances with the state of aging of paper is possible [74]. SPME-GC-MS is highly sensitive and completely non-destructive, but it was found difficult to determine and quantify such marker substances [75, 76].

1.2.5 MIR-Spectroscopy based on direct peak assignment

Molecular vibrations and rotations can be excited by absorption of radiation in the infrared region of electromagnetic spectrum. Many functional groups in organic molecules show characteristic vibrations which correspond to absorption bands in defined regions of the IR-spectrum. These molecular vibrations are localized within the functional groups and do not extend over the rest of the molecule. The mid infrared spectrum (MIR) extends from 4000 to 400 cm^{-1} and is characterized by two main regions [77]:

- Above 1500 cm^{-1}, individual functional groups can be assigned to absorption bands
- Below 1500 cm^{-1} many bands are contained that characterize the molecule. This region is known as "fingerprint region"

When it comes to the detection and characterization of chemical structures, IR-spectroscopy is a very powerful instrument. IR-spectroscopy is very attractive for paper analysis because due to attenuated total reflection (ATR) technique no further sample preparation is needed. Measurements may be taken directly on the paper surface without any detrimental impact on the object under investigation. Measurement time is also very short. One drawback might be that penetration depth of these ATR crystals was found to quite low compared to other IR techniques [78]. When performing quantitative analysis of IR spectroscopy data, Beer's law has to hold true describing a linear relationship between the concentration c, the molar extinction coefficient ε within the sample and the cell wall thickness or effective path length d with the absorbance E at a given wavelength λ:

$$E_\lambda = \varepsilon \cdot c \cdot d$$

ATR-FTIR spectroscopy follows Beer's law as well. Surface densities obtained from ATR absorbance measurement will be influenced greatly by the optical properties of the ATR element, substrate film, adsorbed species and solution that govern the effective path lengths and depths of penetration. A further complication arises when a flat plate ATR element is used, as the effective path length for incident radiation parallel and perpendicular to the plate are not the same [79].

There is a broad variety of applications for MIR applied in the field of pulp and paper analysis. Among them, to name but a few, are studies about cellulose in general, aging, lignin, cross linking and paper treatments [80]. Especially when paper aging [81] was investigated, IR-spectroscopy is a popular tool, covering different aspects like degradation caused by transition metal ions [82] and wet-dry interfaces [83]. An oxidation index of paper defined as a ratio of integrals of bands at 1730 cm^{-1} to that at 1620 cm^{-1} has proved amenable to follow the degradation of cellulose aged under various conditions [84]. The photo-oxidation sensitivity of deacidified papers was studied. By means of infrared, mechanical and physical analysis, it was suggested that the presence of basic compounds enhances the photo-yellowing [85]. It was also suggested to use it as a non-destructive tool for quality control in mass deacidification [86].

It has to be kept in mind that absorption bands will change due to different sampling techniques and that overlapping of bands will complicate accurate assignment. Therefore reference spectra of known substances like pure cellulose, sizing or filling material should be acquired. Without measuring the absolute extent of molecular weight and oxidized cellulose functionalities the degree of deterioration of investigated papers is hard to evaluate [87]. An additional problem with the detection of functional cellulose groups arises from the hygroscopic nature of paper and the sensitivity towards O-H bonds of infrared spectroscopy. Within the paper web different types of water, free and bonded, exist, being able to modify the carbonyl group functionalities by forming IR-innocent hydrates.

1.2.6 NIR-spectroscopy and multivariate calibration

In the paper industry, the demand for non-destructive and fast pulp and paper testing has been met by near infrared (NIR) spectroscopy combined with multivariate calibration [88-91]. NIR spectroscopy it is completely non-destructive and may be applied on originals without a need for sample preparation. It is an attractive analytical option for studies in paper conservation. Being an indirect method, reference methods are necessary to obtain the reference values for calibration. These reference methods may be time consuming, expensive or not applicable to the material discussed above. They will thus be used only for calibration, and will then be replaced by NIR spectroscopy which provides the same parameters in a fast, inexpensive and non-destructive way.

NIR spectroscopy is based on vibrational properties of a molecule and covers the region of 4000 and 12500 cm^{-1} adjacent to the mid infrared region. For a diatomic molecule a harmonic oscillator will follow

$$E_{VIB} = h_v(v+1/2)$$

with v being the vibrational quantum that can only have integer values like 0, 1, 2 and so forth called energy levels. In nature, this model does not hold true and molecules behave rather like anharmonic oscillators. This anharmonicity will lead to overtones ($\Delta v_i > 1$) and combination vibrations (for which several $\Delta v_i \neq 1$) that are much weaker. The intensities of the overtones depend on anharmonicity [92].

NIR is sensitive towards water in the sample. Therefore, one of the main applications of NIR spectroscopy is the determination of water content in food samples. If quantitative analysis is the aim of NIR spectroscopy, variations in moisture content are considered to be acceptable which is of great advantage with the hygroscopic material cellulose [92]. Consequently, sample preparation in NIR is less problematic, but care has to be taken with temperature differences as NIR easily responds to them [92]. Additionally, evaluation of NIR spectra is very difficult, because there will be no single peaks that may be assigned to a defined functional group. The NIR spectra are characterized by rather broad bands originating from overtones and combinations of fundamental vibrations. Nevertheless, the spectra can be explained by the model of anharmonicity and therefore, not only qualitative, but also quantitative analysis is achievable. According to the model of anharmonicity, –COOH, -OH and –C=O are NIR-sensitive functionalities, and a specific response of polymers containing these groups is to be expected [92]. NIR-spectra were even reported to yield information on molar mass and viscosimetry of a polymer [93].

For the interpretation of the large amounts of data (= wavelengths) supplied by one single spectral measurement, data reduction techniques have to be applied in order to extract hidden information within the data matrix. Thereafter, information has to be correlated to a property of interest, which has been determined before by reference methods. A very powerful tool for information reduction and interpretation is multivariate calibration [94]. Multivariate calibration belongs to the field of chemometrics that has been defined as "the entire process whereby data are transformed into information used for decision making" [95]. For the interpretation of NIR spectra, partial least squares (PLS) algorithm has been found to be the most useful one. Usually, data matrices are designed to keep the amount of samples and of variables in equilibrium, but for the multivariate technique PLS 10 to 50 times more variables than samples available can be handled successfully as long as there is sufficient correlation between samples and variable. As strong correlation between single wavelengths is characteristic for infrared spectra, even more variables may be used.

Chemometric approaches are mainly based on six steps. These are:
1. Defining the problem
2. Recording of spectra
3. Recording of relevant data (reference data)
4. Pre-processing of spectra
5. Evaluating the data
6. Calibration, validation and prediction

The definition of the problem was dealt with in the introduction while recording of spectra and reference data will be addressed to in the Materials and Methods section. In the following, the concepts of pre-processing, evaluation, calibration, validation and prediction will be introduced.

General concept

The idea of multivariate calibration is to relate multiple responses from an instrument to a property of interest. As opposed to that, commonly known and applied univariate calibration will relate one single instrument response to a desired property [95]. There are different ways of multivariate sample modelling. All of them have in common that two matrices exist, named X-matrix and Y-matrix. During calibration procedure a correlation between these two data matrices is searched for by mathematical algorithms. Once the correlation between X and Y has been established, the resulting model can be used for future prediction of unknown samples. To build this model, first of all known data sets X and Y have to be generated. They are then used for the multivariate regression model which can than be used for the evaluation of new Y-data by new X-data. Provided this model has been validated, it can replace time consuming measurements of Y-data just by measuring easy to achieve X-data. Very popular examples are infrared spectra used as X-data that will substitute Y-data usually gained by wet chemical analysis. The main goal is to replace the expensive wet chemical analysis by relatively cheap and fast spectroscopy.

Pre-processing of spectra

Pre-processing of spectra is in many cases considered to be necessary to reduce, eliminate or standardise effects on the spectral data without influencing the spectroscopic information. Sources of spectral variations that may influence the acquisition of paper surface spectra are for example, interactions of the compounds caused by intermolecular hydrogen bonds, light scattering from solid samples and poor reproducibility in the measurement itself like path length variations. Other distortions may arise from the spectrometer hardware such as baseline drifts, wavelength shifts, effects from detector non-linearity or stray light and noise originated from the detector, amplifier or analogue-digital converter [92]. Even though pre-processing may improve spectral information it has to be kept in mind that all kinds of data treatment can just as well remove important information [95].

Evaluation of data

Before starting the modelling procedure it is important to get an overview of the data structure to detect outliers or samples that influence the model disproportionately high. This can be achieved by applying pattern recognition techniques. A very important algorithm is called principle components analysis (PCA). For PCA, data are projected onto vectors in an n-dimensional space corresponding to their main direction of variance. The vector that covers most variance is than called first principle component. Following principle components describe the remaining variance within the sample set always searching for an orthogonal vector. Due to the fact that vectors describing the variance are all orthogonal to each other, the original data loose their collinearity which might be useful for some applications. When all variance is finally described no information on the data set will be lost, but unfortunately noise will be modelled as well. As the first principle components explain most variance within the system, the dimensionality is significantly reduced. The most important principle components can be plotted in a 2-dimensional space for data evaluation.

Calibration and validation

The starting point of every multivariate calibration is a so called calibration set that is used to build a model. This calibration set is of outmost importance for multivariate calibration. It should be representative for all samples that are intended to be measured in the future considering comparability of measurement conditions and expected range of results. Within the expected range, sample material should be distributed evenly. As multivariate calibration is very efficiently modelling complex situations, the use of mock-up material is not considered to give any advantage over realistic sampling conditions of original material. The variance found in this original material is translated into information in a chemometric approach.

The model that has been build using the calibration set is then tested using a so called test set. A test set, preferably several test sets, have to correspond to the samples that have to be measured in future applications. Also for the test set(s), X and Y data has/ have to be known in order to evaluate the success, *i.e.* the prediction ability, of the model. The idea of calibration consists in determining the Y-data by the calibration model and than compare them to the known values of X-data. X- and Y-data should be modelled preferably with low residuals; nevertheless the fact of having a good model does not necessarily imply good prediction ability.

When the amount of available data is limited, cross validation might be used as a substitute, but test set calibration is preferred.

PLS is an algorithm for multivariate calibration. It is a method that can be used without explicitly selecting variables. This is accomplished by transforming the measured variables (*e.g.* absorbance values at many wavelengths) into new variables (often referred to as factors) that are used the matrix calculations [95]. This type of Inverse Linear Regression (ILS) methods enables to predict the concentration of one component even if additional chemical and physical sources of variation are present. Implicit modelling means that variation is captured without explicit identification of the variable causing the variation. All that is required is that sufficient variation in the level of the variable be present in the sample [95]. To ensure this natural design is a good choice. This is when many samples are collected over a period of time until one has confidence that variation has been adequately represented. One rule of thumb is to have at least three times as many samples as the expected rank of the system [95]. The approach, therefore, is to choose samples with varying concentrations and measure the spectra on these samples. This is opposite of the classical approach where the independent variable (x) is set and dependent variable (y) is measured [95]. Region selection might additionally reduce variations within one class because often only noise or irrelevant data are obtained in redundant wavelengths [95]. Usually the models become easier to interpret. Even if too many wavelengths have been excluded, PLS is still capable of calculating a reliable model. However, more PLS components are needed to compensate lacking information from the wavelengths. This increases the danger of over fitting. Some features of PLS are listed in table 4.

Table 4. Advantages and disadvantages of PLS according to [95].

Topic	Disadvantages	Advantages
Theory	Difficult theoretical background leading to "black-box-feeling" when working with this method	
Amount of samples	When relying on "natural design" relatively much samples are needed for model construction	Significant sources of variation can be accounted for without knowing their origins
Modelling	The determination of how many factors to use in the model is not straightforward	
Model diagnostics	Experience is needed to apply model diagnostics and validation properly	When familiar with model diagnostics they are a powerful tool to asses model capability

PLS is considered to be most useful for the evaluation of spectra and shall therefore be described more in detail here. The main difference between other multivariate approaches consists in the fact that for PLS the influence between both data matrices will be taken into consideration for calibration. There are two PLS-algorithms, PLS1 and PLS2, their main difference being that PLS2 can handle several Y-variable simultaneously, while PLS1 is operating with one Y-variable only.

Information that are relevant for Y should be concentrated in the first components, but when dealing with complex problems even more than ten components might be needed to describe the model in a sufficient way. Surface of the sample material might be one external influence considerably increasing the complexness of a model.

A first step toward the understanding of PLS might be the idea that a parallel PCA is performed on both matrices, X and Y. The crucial point is that these PCAs are not run independently, but influencing each other. PLS links the two data sets to each other creating a u-score vector in the X-matrix that is the starting point of a t-score vector in the Y-matrix. The t-vector will then again be used for the X-data matrix. PLS-algorithm is based on the mutual exchange of $u_1 \rightarrow t_1$ and $t_1 \rightarrow u_1$ till convergence. Both data spaces will be modelled in dependence of each other. Balancing X- and Y-data, PLS reduces the influence of big X-variations that do not correlate to Y. PLS will lead to two data sets: they are called loadings, P, and loading weights, W.

In order to further apply the models derived from PLS-algorithms, validation has to be done to check the quality of the model. Internal validation is called cross-validation; it uses samples for validation that have originally been used for model set-up. Cross validation is used when there are not enough samples available to set aside a calibration and a test set. As for this type of validation the number of samples equals the number of sub models that are produced by sorting out one sample, it is also called "leave one out" calibration. The left out sample is then predicted by the sub model.

For external validation, samples are used that have not been part of the model building process. This type of validation is also called test set validation.

Both types of validation compare the predicted matrix (Y_{pred}) to the reference matrix (Y_{ref}). Each validation process yields a measure for the prediction error, *i.e.* the error that has to be expected when using the model for new predictions. Additionally, a modelling error will be calculated, too. In general, the prediction ability is best, when the root mean square of prediction (RMSEP) is lowest. In order to get a realistic idea about how big the RMSEP should be, the user has to know the standard deviation that is typical for the chosen reference method. As a rule of thumb, RMSEP will be two or three times higher than standard deviation as the errors of the model will add on errors of the reference method. Modelling error is calculated by building a model on the basis of the two calibration matrices X_{cal} and Y_{cal}. At this point the correct amount of components has to be assumed. Data for X_{cal} will be inserted into the model to predict $\hat{y}_{cal}$:

$$X_{cal} + model \rightarrow \hat{y}_{cal}$$

When comparing predicted values to measured, the modelling value for a certain amount of components is obtained:

$$\text{modelling error} = \hat{y}_{cal} - y_{cal}$$

This calculation is performed for all objects. When squared differences are summed up and the average for n objects is calculated, the residual Y-variance of calibration is obtained:

$$\text{residual variance}_{cal} = \frac{\sum (\hat{y}_{cal} - y_{cal})^2}{n}$$

The value is expressed as root mean square error of calibration (RMSEC). Values are given in absolute numbers.

The root mean square error of prediction (RMSEP) is calculated in a similar way, instead of y the test set, that has not been used for modelling is inserted:

$$\text{residual variance}_{val} = \frac{\sum (\hat{y}_{val} - y_{val})^2}{n}$$

The smaller the difference between these two values, RMSEC and RMSEP, the better the model is. The more components of the PLS model are used, the lower the difference between the values will be, but an optimal amount of components has to be figured out. When choosing a too low amount of components, the modelling is not sufficient for further predictions. This phenomenon is called "under fitting". Using too many components will improve the fit of the present model, but decreases prediction ability for further samples. This is called "over fitting". Generally, a low amount of components is preferred.

Prediction

After thorough testing of the model it can be used for prediction of unknown sample material. Besides outlier warnings and prediction uncertainty limits usually incorporated in the used software, there are no means to actually test prediction ability once a model has been chosen. Freshly predicted samples should then have the same errors like during validation procedure with known samples.

During PLS predictions, new X data are projected onto the chosen amount of components. Y is then estimated according to the projected scores und loading matrices, T and P. The well known regression equation:

$$Y = b_0 + b_1 x_1 + b_2 x_2 + \cdots + b_n x_n$$

is transformed into:

$$B = W(P^T W)^{-1} Q^T$$

for the prediction within a PLS-model.

It is recommended to use the model for predicting values that have been analysed by the reference method to further check the prediction quality and control the validity of the model. Again, predicted values are plotted against measured ones and the differences between these two values are compared to each other.

1.3 Relevant topics in paper conservation

As already outlined in chapter 1.1, historic papers are difficult to evaluate in terms of their actual condition and future perspectives. Chapter 1.2 introduced key parameters of cellulose, the main constituent of paper, and analytical approaches to determine them. The following chapter will present some of the relevant research topics in paper conservation that deal with condition rating and degradation phenomena caused by oxidative and hydrolytic processes.

1.3.1 Condition rating of historical papers

Condition rating in the context of paper conservation is difficult to achieve, mainly because the original condition of the papers under investigation is unknown. The same holds true for the storing conditions of these papers. As long as original quality directly after production and environmental conditions for storing are unclear the definition of the present condition remains impossible. The change of a certain paper in relation to its original condition cannot be estimated. The only option in this context is the comparison of papers having the same age, same history, same fibre composition and same production background. Even this is complicated as only assumptions on this common history are possible in most cases.

Basically, paper can be stored as a single document, laying or standing in a box or envelope or bound in a book usually standing in a shelf. Access of light, oxygen, and other gaseous substances is different for both forms of storage, so that the question arises how storage influences aging parameters and if there are any detectable differences between those two types of storage. Library and archival material is hardly ever stored as a single sheet but rather in convolutes formed by book bindings or packages of paper sheets. On the one hand, paper inside of a book or in the middle of a convolute is hardly ever prone to light damage and also the access of oxygen and external pollutants is limited, but on the other hand volatile degradation products will accumulate especially in the centre of bound material or in a convolute. Until now, some investigations exist about the strength distribution of pages throughout whole books divided into centre and outside pages [96] and about migration of volatile degradation products throughout stacked paper simulating stored books [97, 98]. Both studies indicate that within one edition of a book fairly different paper conditions have to be expected due to the migration of degradation products. Contrary to that little is known about the course of acid formation, cellulose chain shortening and oxidation within a single sheet of paper.

From the analytical point of view, paper degradation can be monitored by measuring pH and molecular weight to determine the degree of acid hydrolysis, and carbonyl plus carboxyl group content to estimate oxidation phenomena.

1.3.2 Paper and transition metal ions

General

There are many investigations dealing with the oxidative potential of transition metal ions on cellulose because frequently used irongall ink and brilliantly coloured copper acetate are known to cause damage on works of art on paper. Cellulose in paper is mainly attacked, but also protein based constituents like animal glue and gelatine or parchment as a writing support will suffer from accelerated deterioration.

Visually, signs of degradation can be perceived by a discolouration of the pigment or ink spreading out on the surrounding paper, as well as by a loss of mechanical strength. Although parallels do exist between corrosion of cellulose caused by copper containing pigments or by ferrous and ferric ions (so-called irongall ink corrosion) they are not following an identical degradation process [99, 100]. The influence of these writing and painting media on paper leads to differences in the behaviour towards water. Humidification will no longer occur homogenously (figure 11). This observation is eventually attributed to migration of ingredients from writing or painting media into the surrounding paper material. The water required for migration processes might be provided from paper itself, from ambient humidity or accidental wetting.

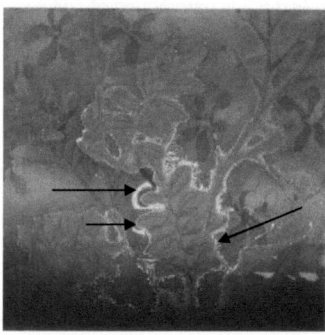 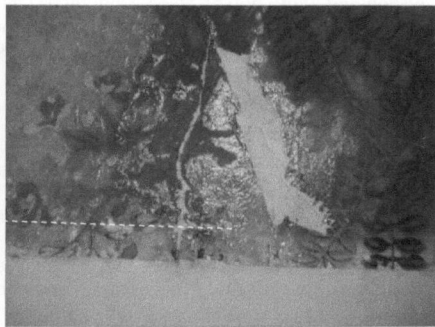

Figure 11. Example of paper degradation caused by a copper containing pigment (pictures taken by Aram Ter-Akopow). Left: Water will penetrate the paper more easily in areas close to the areas with green pigment (black arrows). Right: When the painted paper was covered to protect it from the access of light (below white line), no colour changes in the pigments and the paper occurred.

The influence of selected metal traces on the colour and colour stability of purified cotton linters has been investigated [101]. In the 1970's Emery and Schroeder investigated the oxidation of cellulose and hemicelluloses in the presence of iron and found that, contrary to other reports, the reaction was favoured in acidic conditions [102]. Especially the negative influence of iron and copper ions on the permanence of paper has been worked on extensively, emphasizing the catalytic role of even trace amounts of different transition metal ions [103-106]. Recent research extended on the catalytic activity of Cd(II), Co(II), Cr(III), Mn(II), Ni(II), and Zn(II) in addition to Cu(II) and Fe(III). At neutral pH, Cu(II) was found to be the most active cation, while Cd(II) and Zn(II) did not exhibit any catalytic activity. Small amounts of Zn(II), Co(II) and Mn(II) will exhibit pro-oxidative effects [104].

Historic irongall ink

Historic irongall ink is typically composed of two main ingredients, iron(II)sulphate and a vegetable tanning agent, mostly extracts from gall apples. The extractives of gall apples contain tannins, which will form a deeply coloured complex with iron(III) that is known as irongall ink (figure 12). Upon complex formation, sulphuric acid is produced.

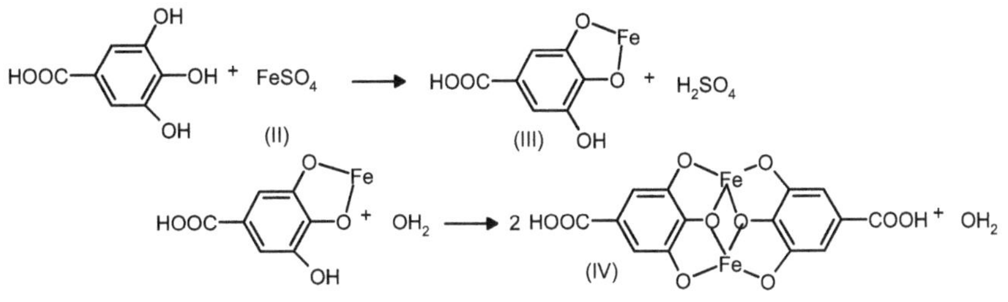

Figure 12. Colour forming complex in irongall ink. Aqueous gallic acid (I) and iron(II)sulphate (II) form a colourless, water soluble Fe(II)gallic acid complex (III). Oxidation will then lead to the black irongallate complex (IV) [107].

Depending on source and availability, the historic iron sulphate also contained impurities such as copper, zinc, manganese, aluminium and magnesium salts, establishing a complex system of different stabilizing and destabilizing metal ions [108]. Tannic acid is a polyphenolic substance. This class of substances was investigated mostly in the context of food chemistry. Tea polyphenols were found to be powerful antioxidants [109, 110]. The pro-oxidative effects of gallic acid in Fenton-type systems containing H_2O_2 and Fe-(III) at different pH and temperature have been addressed to. The overall effect was found to be dependent on the ratio of gallic acid and Fe(III) in the reaction medium [111]. Ideally, iron ions should be bound in colour forming complexes provided there are enough chelating agents, i.e. tannic acid available. Due to the inhomogeneous source material this cannot be assumed, therefore, free transition metal ions are expected to act in redox reactions leading also to Fenton type radical formation [102, 112, 113]. As the stability of irongall inks relays on the ratio between iron ions and tanning agent the concentration of tannic and gallic acid in the chosen supply is of great importance. In natural sources like gall apples which have traditionally been used, the concentration of these components vary considerably, even within one species and upon aging [114]. Additionally, all kinds of modifiers have been added to irongall inks, usually gum Arabic as protective colloid and alcoholic liquids to improve extraction from gall apples, but later also sulphuric acid [115, 116].

Copper acetate pigment

Copper acetate is one of the oldest synthetic pigments. Traditionally it was produced by a reaction between metallic copper and acetic acid, either deliberately added or freed during degradation of organic matter added to a copper plate, with an excess of oxygen and carbon dioxide. Several additives have been used like urine, salt, honey and wine. In historical preparations for painting purposes, copper acetate-based pigments are a mixture of neutral and basic copper acetates, which are partly water soluble from the beginning. The binding agent was mostly glue or gum Arabic. It has been known for a long time that this pigment would harm its support material; therefore glue has always been suggested as a protective colloid. A more comprehensive overview about copper acetate preparation for painting purposes is found in [117, 118].

Approaches for analysis and identification of underlying mechanism

Like historic copper acetate, historic irongall inks are a hardly reproducible system which makes studies in this topic quite difficult to handle. In case of irongall ink, model papers and naturally aged papers have a common degradation feature when it comes to their main migration characteristics. In both cases, it is the enhanced mobility of sulphate ions whereas iron ions mainly remain in their original position [119]. It was also found that typical migration behaviour of naturally aged inks can be simulated using cycling humidity aging [120]. During natural aging the degradation of the binding agent - mostly gelatine or gum Arabic - is accelerated as well, promoting the release of ions into surrounding paper. During this migration process, cellulose is chemically altered and deteriorated or even fully destroyed [20].

Considering all the ingredients of transition metal containing inks and pigments, the question arises what the driving mechanism behind the degradation they cause on paper is. Increasing emphasis was put on the conclusion that irongall ink corrosion seems to be a system with two contributing effects: acid hydrolysis due to sulphuric acid that is liberated during the formation of the irongall complex and additionally charged to improve writing properties, and transition metal ion catalyzed autoxidation [120]. There is not much literature explicitly addressing the analysis of oxidative or hydrolytic reaction pathways in the irongall ink system and the contribution of the single ink components to the degrading action. On paper, the volatile organic compounds (VOC) of different irongall ink ingredients, separate and combined, have been analyzed. Tannic acid and gum Arabic alone do not produce VOCs, while their production is observed when only sulphuric acid and ferrous sulphate are present. Most VOCs are detected when irongall ink as a combination of both substances has been used [121]. This may be regarded as an indirect proof of synergistic effects in the irongall ink corrosion system.

Treatments

In the context of oxidative bleaching in paper industry troubles arising from metal content in pulps have been answered by the addition of magnesium compounds. It will slow down the catalytic oxidation triggered by transition metal ions and helps to improve pulp quality [122, 123]. In contrast to pulp and paper industry manipulating with slurries, the distribution of magnesium compounds in a solid paper matrix is a lot more complex and hardly ever achieved. Additionally, strong alkalinity of magnesium salts is suspected to cause extra degradation on heavily oxidized paper due to β-elimination.

According to the synergistic model, acid hydrolysis caused by sulphuric acid and autoxidation due to iron ions degrades paper objects of historical and/ or esthetical value. Treatment options should therefore inhibit the two different degradation mechanisms. Translated into conservation treatment of objects that suffer from irongall ink corrosion, a deacidification and a chelating step is needed for successful treatment. This approach has been investigated in depth [112]. Many research works dealing with the treatment of irongall ink model papers has been performed to further establish the beneficial use of the calcium phytate/ calcium hydrogen carbonate treatment [119, 124].

Additionally the influence of gelatine sizing was investigated as proteins are reported to possess cation binding capacity that should help to immobilize metal ions and keep them from reacting [125, 126]. Gelatine is believed to slow down irongall ink corrosion as it reduces ion mobility and, good penetration provided, forms a protective layer between inks or pigments and paper carrier. Being a protein, gelatine may also show a stabilizing effect *per se* when applied to irongall ink containing paper. A possible complexation of metal ions by gelatine is discussed in that respect. Also, gelatine is often used as a resizing agent as part of the conservation treatment because parts of the original sizing of paper are assumed to get lost during aqueous treatments.

1.3.3 Bleaching in the restoration context

Bleaching is defined a chemical process aimed at the removal of colour in pulps derived from residual lignin or other coloured impurities such as resin compound and dirt originating from wood, the cooking process or from external sources [127]. The bleaching process in industrial pulp production is necessary for optical reasons. While in native wood, lignin is only slightly coloured, it forms darker reaction products upon pulping. The complete removal of these coloured products via prolonged cooking is not recommended as it leads to enhanced carbohydrate degradation. Brightness gain through bleaching is slowed down at higher brightness levels.

Brightness is measured according to ISO standard method by measuring the reflection of blue light at 457 nm from an opaque amount of pulp or paper sheets. According to this scale, 100 % reflection would correspond to a perfectly white sheet. In pulp industry about 95 % ISO brightness can be obtained, values > 88 % ISO brightness are considered to be fully bleached pulps [127].

Chemicals used for industrial bleaching have been divided into three groups describing their reactivity in terms of electrophilic or nucleophilic attack, pH conditions, their reactivity towards lignin and carbohydrate structures. Typical chemicals are oxygen, chlorine dioxide, ozone and hydrogen peroxide. With the exception of a limited use of sodium borohydride in a stabilizing step after ozone bleaching and sodium dithionite in bleaching of mechanical pulp, most of the bleaching chemistry in pulp production corresponds to oxidation reactions. Compared to chlorine dioxide, oxygen is not very selective and will also attack native carbohydrate structures, *i.e.* cellulose and hemicelluloses when highly reactive radicals are formed under bleaching conditions.

Paper conservators use bleaching treatments to increase overall brightness or to reduce local spots that visually distort the observer, especially when dealing with graphic art on paper. Conservators are aware of the detrimental effects on cellulose caused by bleaching, and therefore it is not considered to be a conservation (= aiming at the improvement of paper permanence), but a purely esthetical treatment. Common bleaching agents used by conservators are hydrogen peroxide (either as commercially available solution or as in situ produced upon light bleaching), calcium hypochlorite, sodium borohydride and potassium permanganate [128]. The later one is not very much favoured any longer, while light bleaching has experienced increasing interest [129]. While the assumed detrimental effects of oxidative bleaching have lead to a very restricted used, reduction treatments are considered to be a somewhat mild alternative, even beneficial effects have been reported [130, 131].

However, in the context of this work, studies of restoration bleaching treatments have not been investigated and shall be addressed to in future research.

1.3.4 Wet-Dry interfaces

When dealing with old papers stains originating from the influence of water are commonly encountered. Local treatments and surface testing methods for pH or for the presence of certain ions also request the local application of small quantities of water. Often, these interactions will lead to the formation of a coloured line at the wet-dry interfaces. As a certain amount of dirt and degradation products may reasonably be assumed in historic papers, these coloured lines have often been referred to as accumulated decomposition products and soil by capillary action.

In contrast to that assumption, an interesting observation can be made when pure and clean sheets of cellulose are inserted into water and due to capillary forces the water is soaked up to a certain height where it will stop and form a wet-dry interface. This interface is easy to see because it is usually accompanied by the formation of a yellow to brown line. Additionally, fluorescence of this interface can be observed under UV-irradiation at 254 nm. This phenomenon is more pronounced when the chromatography process is extended in time, but even after applying a droplet of water on a paper surface and immediate removal of water fluorescence will be observed.

The changes in fluorescence intensity of a cotton cellulose paper without additives in which a wet-dry interface has been created have been characterised spectroscopically. It was found that the intensities of the MIR spectra recorded at the wet-dry boundary were approximately three times higher than those recorded on the original paper [132].

Oxidative degradation has been detected by methylene blue absorption on the interface. Next to oxidation also degradation and acidity of the coloured region have been reported.

Fungal growth is favoured on the tideline region. Several reasons have been discussed to describe the formation of wet dry interfaces, also called tidelines [133]:

- A restricted dispersion of free or slightly aggregated cellulose molecules by water and their transfer by convection to the tideline region
- Modification of the cellulose over the whole area of the test strip, and the collection of the modification products occurring in the wet part by convection at the boundary line.
- Modification by water or by water and air, in the wet region only, and transport of the modified product to the boundary line.
- Initial presence or formation during the experiment in the wet area of a substance capable of reacting with the cellulose. Subsequent accumulation of much of this substance at the boundary line before it has been able to react with the cellulose in the wet region generally.
- A specific change in the cellulose which can take place in the conditions which exist in the boundary region and nowhere else.

More recently, results from a study of wet-dry interface of pure cellulose Whatman filter paper were presented. According to this publication, degradation products in the tideline region were formed by oxidation. Fourier transform infrared spectroscopy spectra showed the presence of carbonyl groups from aldehydes and ketones. Thin layer chromatography and gas chromatography coupled with mass spectroscopy (GC/MS) could identify sugars and uronic acids in the degradation products extracted from the cellulose [83].

Other authors have additionally studied the peroxide content that also indicated oxidative degradation within the tideline region [134].

1.3.5 Mass deacidification

The predominant aging pathway of cellulose and therefore of paper is hydrolytic chain scission closely related to lowered pH [11, 18, 21, 135]. Purely hydrolytic scission chemistry is described by the formation of one additional carbonyl group per chain scission, with no significant increase in carboxyl groups [135]. During the investigation of old, well preserved books it was found that their pH was mostly only slightly acid or even alkaline. The reason for this was some content of magnesium and calcium ions in historic papers to establish moderate pH conditions [18]. It was therefore concluded that an active input of alkaline substances should improve paper permanence.

Early applications of deacidification included over-night soaking in solutions of magnesium and calcium salts [18]. In the course of the years, these treatments have entered conservation workshops and are by now performed on a more routine base. During the immersion in the solution, water soluble degradation products will be washed out of the paper while a surplus of alkaline substance will be deposited into the paper web serving as an alkaline reserve against future acid uptake or development. In today's aqueous deacidification treatments, solutions of $Ca(HCO_3)_2$, $Mg(HCO_3)_2$ and $Ca(OH)_2$ are used, all of them realizing a different pH in aqueous solution. The approach towards treatments has changed since its first suggestions: immersion times grew considerably shorter and the pH of the paper to be treated is monitored before and after the treatment, while the pH of the immersion bath is also controlled. The adverse effects of a too alkaline pH have also been recognized.

With regard of the huge quantities of acidified books, even large scale mass treatments for the deacidification of whole library and archive stocks have been invented and successfully introduced into the market. With few exceptions, mass deacidification is performed on the basis of magnesium containing components; many of the systems use non-aqueous solvent systems in order to be able to treat whole books without damaging the bindings through the swelling action of water [136].

The following key questions inflicting aqueous single item treatments as well as non-aqueous mass treatments have been asked in the context of deacidification:
- Which substance should be applied for deacidification?
- How much of an alkaline reserve should be deposited in the paper?
- How does homogeneity of the objects to be treated and the treatment itself influence the result?
- Will inhomogeneous deacidification have any negative implications on further aging?
- How should the success of deacidification be monitored?

Especially the use of magnesium compounds is also criticized [137]. They are suspected to cause yellowing at least in lignin containing papers [138], and due to their higher alkalinity, depolymerization of cellulose was also reported [139]. This is even more important for non-aqueous mass deacidification when substances are used that have the potential for an even higher pH to be formed intermediately than aqueous $Mg(HCO_3)_2$ solutions. Nevertheless, in other investigations about the effects of aqueous deacidification on historic paper samples using $Mg(HCO_3)_2$ solution, it was found that even better results as for deacidification using $Ca(HCO_3)_2$ solutions were obtained, probably because pH will play a less important role in sized papers with additives than in Whatman filter paper [140].

When it comes to the question how much alkaline reserve should be deposited into the paper web some suggestions require up to an equivalent of $CaCO_3$ of 2%, translated into 1.7% of $MgCO_3$ [141]. A higher alkaline reserve can obviously neutralize more acids formed in the future, but it will detrimentally effect paper smoothness and encourage uptake of gaseous pollutants [27]. Therefore, the efficiency of lower amounts of alkaline reserve has also been addressed to [142].

Inhomogeneous degradation processes caused by differing access of light and oxygen or different rates of ventilation to let volatile degradation products escape from the inside of a book might be one explanation for the observation that in few books some areas of a sheet are perfectly deacidified while other areas remain acid.

Meaningful tests for the evaluation of mass deacidification treatments are still missing. Tests that are applied today for quality control include on the one hand visual control for unwanted effects on the optical appearance of the treated material and on the other hand on pH, amount and distribution of alkaline reserve to document the success of deacidification, colour measurements, and several mechanical tests that are performed on mock-up material due to their high sample demand. Non-destructive approaches are still under development. Detailed analysis of the impact of deacidification on the molecular structure of cellulose is still missing. Most investigations about the added benefit of this treatment have been based on mechanical tests which are far too insensitive to monitor changes under artificial aging conditions [143].

2 Objectives

The present work will focus on the application of fluorescence labelling followed by gel permeation chromatography for the analysis of cellulose and paper samples (figure 13). The two techniques, in the following named CCOA-analysis for carbonyl group determination and FDAM-analysis for carboxyl group determination were developed for the analysis of pulp, paper and fibre properties and yields highly sensitive information about oxidation and chain scission in cellulose. These features make the fluorescence labelling approach very attractive for the analysis of various phenomena in the context of paper conservation.

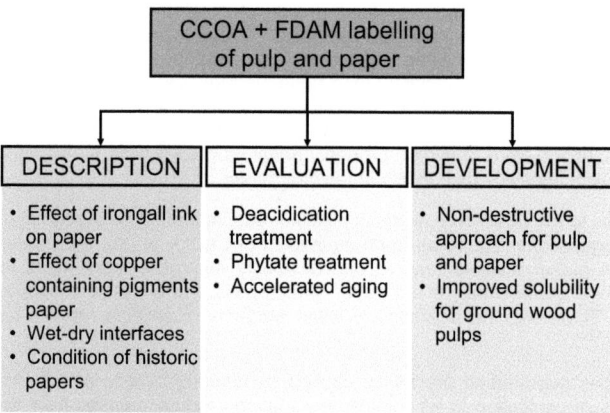

Figure 13. Schematic overview of objectives

Description

Degradation processes of historic papers can be described on basis of wet-chemical analysis using model papers and original sample material. The condition of historic papers is described by the extent of oxidation and chain scission, parameters that are made available by the chosen analytics.

The extent of oxidative action triggered by transition metal ions contained in paper, pigments or inks can be studied using carbonyl and carboxyl group content as well as average molecular weight and molecular weight distribution. In this context the question of spatial distribution of damage in relation to the ink or pigment application is of special interest. When spatial distribution of oxidative damage is looked at then consequently the question arises in how far metal ion migration corresponds to the observed damage.

Tideline formation noticeable under UV-radiation at 254 nm can be observed after all local wetting procedures. A certain extent of spontaneous oxidation is one theory. As oxidation and chain shortening can easily be measured using fluorescence labelling and GPC, this phenomenon was also studied in the framework of the present work to find out more about the mechanisms leading to fluorescence.

Evaluation

Next to the description of damage, opportunities to inhibit or slow down the degradation processes are of great interest. In the context of this work the efficiency of combined calcium phytate/ calcium hydrogen carbonate treatment was investigated. Especially the comparison between sample material and the behaviour of treatment on historic sample material is considered to be of great interest. Concerning treatment strategy it may be interesting to find out if e.g. ink corrosion always follows the same patterns or if there will be differences within this type of damage category. Comparison of historic sample material should reveal similarities and differences within one damage category and between copper and iron induced damage. When comparing model papers that have been artificially aged to induce damage and historic originals then an important question is obviously in how far the natural aging may be simulated within a short period of time using relatively pure sample material.

In the context of paper deacidification usually surface pH measurement and determination of alkaline reserve are performed. Nevertheless, the question in how far acid and oxidative degradation have been stopped should be answered by more detailed analysis, directly related to the cellulose molecule.

Development

All degradation phenomena described above, but especially acid hydrolysis, may be found in ground wood papers from the 19th and 20th century. The success of estimating ground wood sample material depends on the possibility of dissolving them in the DMAc/ LiCl system used for GPC analysis. Therefore emphasis has to be put on improving their solubility.

The correlation of wet chemical analysis to NIR spectra by means of multivariate calibration can be used to predict unknown samples to obtain analytical data without performing wet chemical analysis. This is highly desirable for several questions in the context of paper conservation and condition rating. Multivariate calibration systems can cope very well with internal sources of variation in the sample material and describe theses variation in models.

3 Materials and methods

3.1 Pulps, papers and inks

To study the degradation pathways in detail and in order to test new conservation treatments as well as storage options for historical papers, mostly model papers have been used. Precious originals cannot be employed in studies with unknown results and cannot provide sufficient amounts of sample material for large numbers of screening tests. First of all, the choice of paper making fibres, pigments, binders and preparation techniques is crucial for obtaining authentic model material. A similarity of these factors with regard to historic samples is the prerequisite to conclude from artificial sample material to original, historic papers. Yet, some doubts remain whether artificially aged sample material will follow a similar degradation pathway as historic papers of comparable constitution. While for irongall corrosion a large number of naturally aged material has been investigated in numerous tests, mainly destructive, mechanical ones, the number of studies about historic papers and manuscripts exhibiting degradation induced by copper pigments – so-called "degradation caused by copper ions" is rather limited.

3.1.1 Pulps

Cotton Linters

Cotton linters was obtained from Buckeye[3]. A general description of the MW, the polydispersity index (PDI) and carbonyl group content parameters was performed at BOKU by means of fluorescence labelling followed by GPC as described in chapter 3.5. The results are presented in table 5.

Table 5. Specifications of Cotton linters from Buckeye analyzed at BOKU (n = 17).

Sample name	Mean	SD [absolut]	SD [%]
M_n [kg/mol]	89.4	12.5	13.9
M_w [kg/mol]	138.9	7.8	5.6
M_z [kg/mol]	196.7	10.0	5.1
PDI[1]	1.6	0.2	
C=O [µmol/g]	0.8	0.0	5.0

[1] PDI: Polydispersity index: M_w / M_n

---

[3] Buckeye Technologies Inc., 1001 Tillman, Memphis, TN 38112, phone: (901) 320-8100, fax: (901) 320-8836, email: info@bkitech.com

BKZO

Bleached beech sulphite pulp (BKZO) was obtained from Lenzing AG (www.lenzing.com). The pulp is described in table 6.

Table 6. Specifications of BKZO from Lenzing as provided by the producer.

Parameter	Unit	Limits
Viscosity (Cuen)*	ml/g	typically: 545±75
Ash content	%	max. 0.08
Brightness	%	min. 90.0
Alkali resistance R18	% (m/m)	min. 92.5
Alkali resistance R10	% (m/m)	min. 86.0
Fe	mg/kg	max. 5
CaO	mg/kg	max. 150
SiO$_2$	mg/kg	max. 120

* depending on clients specification

Pulps for NIR-PLS

In total 118 different pulps were obtained from different sources, raw materials and modifications for the development of a non-destructive approach based on near infrared spectroscopy and multivariate data analysis. They were weighed to 2 g portions and lab hand sheets were prepared out of them at Lenzing AG (www.lenzing.com) according to ISO 3688_1999 Pulps - Preparation of laboratory sheets for the measurement of diffuse blue reflectance factor. Wet chemical analysis was performed on samples taken from these hand sheets. Analytical details and a short description of these papers are summarized in a separate lab journal.

3.1.2 Model papers

Whatman filter paper

Test papers were prepared of Whatman filter paper no. 1 containing almost pure α-cellulose without additional sizing or fillers (table 7). Even though it does not entirely reflect the characteristics of rag paper, the changes at the cellulose of Whatman filter paper can be measured without having effects originating from hemicellulose, residual lignin and other naturally occurring paper components (table 1 on page 14). Samples with no ink on it are referred to as reference paper.

Table 7. Description of typical properties of Whatman filter paper no.1 provided by the manufacturer

Particle Retention Liquid	Air Flow Rate	Ash	Typical Thickness	Basis Weight	Wet Burst	Dry Burst	Tensile M/D Dry
(μm)	(s/100 mL/in2)	(%)	(μm)	(g/m2)	(psi)	(psi)	(N/15 mm)
11	10.5	0.06	180	88	0.3	16	39.1

Table 8. Description of typical amounts of trace elements of Whatman filter paper provided by the manufacturer expressed as µg/g

Aluminum	<0.5	Iron	5
Antimony	<0.02	Lead	0.3
Arsenic	<0.02	Magnesium	7
Barium	<1	Manganese	0.06
Boron	1	Mercury	<0.005
Bromine	1	Nitrogen	23
Calcium	185	Potassium	3
Chlorine	130	Silicon	20
Chromium	0.3	Sodium	160
Copper	1.2	Sulfur	15
Fluorine	0.1	Zinc	2.4

To produce wet-dry interfaces on pure cellulose, Whatman filter paper was chosen. A strip of paper was mounted and about 1 cm of the paper at the lower bottom was dipped into demineralized water (figure 16 on page 53). Due to capillary action water was transported upwards for a certain distance till no further transportation took place. A faint brown line marked the area where no further water transportation appeared. One set of sample material that was prepared under lab conditions was obtained from Dr. Anne-Laurence Dupont[4], while another set was produced on the same paper material received from Paris in BOKU under clean room conditions.

Model rag paper (degradation caused by copper ions)

Test papers were prepared of modern handmade paper composed of chemically pulped hemp fibres by Gangolf Ulbricht[5]. During the sheet forming process no additional sizing or fillers were used. These impurities originating from raw material make the paper more similar to original rag papers, while no difficulties upon analysis are expected due to filling or sizing agents. This type of model paper is intended to link filter paper to historic paper.

Model rag paper (irongall ink corrosion)

Test papers were prepared of historic textiles made out of fibres from flax (*linum usitatissimum*) without additional fillers also prepared by Gangolf Ulbricht[5]. This type of fibres is composed of 76-80 % of cellulose, 2-3 % of lignin, of 12-17 % of hemicellulose and 5 % of other components (wax, pectins and fats). These papers were used in the context of a DFG-project[6].

---

[4] Centre de Recherches sur la Conservation de Documents Graphiques (CNRS), 36 rue Geoffroy-Saint-Hilaire 75005 Paris

[5] Gangolf Ulbricht, Mariannenplatz 2 10997 Berlin, Germany, fon/fax: 49.0.30.615 81 55, mail: gangolf.ulbricht@p-soft.de

[6] Deutsche Forschungsgemeinschaft (DFG): „Restaurierung der durch Tintenfraß beschädigten Handschriften des Savigny-Nachlasses. Anwendung der Calciumphytat-Calciumhydrogencarbonat-Behandlung und partieller Stabilisierung in der Praxis", LIS 2 – 557 22 (1) Marburg, INST 1980/1-1, at the Universitätsbibliothek Marburg from 2004 to 2007.

Test paper containing ground wood ("NOVO")

According to "Empfehlungen zur Prüfung des Behandlungserfolges von Entsäuerungsverfahren für säurehaltige Druck- und Schreibpapiere" the success of deacidification treatments should be controlled by the analysis of two test papers, one of them containing ground wood [141]. *NOVO* paper will meet these specifications and is provided by KLUG CONSERVATION (www.klug-conservation.com). In table 9 some facts about *NOVO* are summarized.

Table 9. Specifications of NOVO paper

Parameter	NOVO
Pulps	50-65 % of ground wood, 20-35 % of pulp
Filler	12-15 % of kaolin
Grammage	90 g/m²
Surface	Machine finished
Sizing	Ink penetration 4 min
Sizing agent	Rosin
Surface sizing	None
pH	About 4.5 (alum)
Optical brighteners	None
Wet strength retention	Low

3.1.3 Model inks and pigments

Irongall inks

When ink was applied on any type of the described papers, it was plotted with a Roland DXY-1150 plotter. The general preparation consists in the separate dissolution of the calculated amounts of salts, tannic acid and gum Arabic in demineralized water, their mixing and finally filling the mixture up to 25 mL using demineralized water (table 10).

Table 10. Overview of investigated ink modifications

Ink-Type	Abbreviation	Composition
Balanced ink	OA	Iron(II)sulphate (1.05 g), tannic acid (1.7 g), gum Arabic (0.79 g)
Unbalanced ink	OU	Iron(II)sulphate (1.05 g), tannic acid (1.24 g), gum Arabic (0.79 g)
Copper ink	OK	Iron(II)sulphate (0.998 g), copper sulphate (0.053 g), tannic acid (1.23 g), gum Arabic (0.79 g)
Tannic acid	OT	tannic acid (1.23 g), gum Arabic (0.79 g)
Iron(II)sulphate	OE	Iron(II)sulphate (1.05 g), gum Arabic (0.79 g)

Copper acetate pigment

Copper acetate (90 % basic and 10 % neutral copper acetate) was printed on unsized model rag paper according to historical protocols. The pigment was bound in skin glue media. The pigment was provided and printed by Lutz Walter[7]. No detailed information about the ratios of single ingredients are available, the mixture was produced to meet the purpose of wooden model printing.

3.1.4 Historic papers

Test papers (20th century), "T"-samples

Original papers from the 20th century were obtained from Preservation Academy GmbH Leipzig (PAL). They were cut out of historic books owned by PAL for test purposes. Surface pH data were partly supplied by PAL.

Historic papers (16th to 19th century)

More than 300 historic rag papers were obtained from Prof. Dipl.-Ing. Dr. Gerhard Banik (State Academy of Fine Arts Stuttgart) for the development of a non-destructive approach to establish condition rating based on near infrared spectroscopy and multivariate data analysis. Analytical details and a short description of these papers are summarized in a separate lab journal. The data obtained from these papers were also used for a general statistic description of historic rag papers.

Historic letters containing irongall ink

Paper samples containing irongall ink are rather widely available compared to, e.g., historic samples with degradation caused by copper ions. A number of letters from the 18th and 19th century with irongall ink writing on paper from unknown background was purchased via internet and analyzed. Three examples are given to illustrate different type of naturally occurring ink corrosion on paper. They are referred to as no. 296, no. 298 and no. 300. The papers were analyzed without further treatment or aging to study aging and degradation processes.

"K"-samples containing irongall ink

A set of similar looking paper material was purchased in the framework of a DFG-research project[8] to have original irongall ink sample material. This set of naturally aged papers with irongall ink was used for testing the phytate treatment on original paper. They are referred to as "K" samples.

Historic material containing copper pigment ("Burgerbibliothek")

Loose fragments from the prayer book of Ursula Begerin, Codex 801 of the former Burger Library Bern (today Zentrum historische Bestände, Universitätsbiblithek, Bern, Switzerland), have been analyzed. Codex 801 dates back to 1390 – 1494. It contains handwritten prayers and paintings from the *Reuerinnen* convent Strasbourg on 195 pages (140 x 90 - 95 mm) with a leather-bound wooden cover. The rag paper appeared rather thick to the grip, but was apparently weakened at areas covered with green pigment. In contrast to the typical appearance of degradation caused by copper ions phenomena that is characterized by brownish discolorations of the pigment and paper (see figure 12), the paper of the codex appeared flawless white, also at the cross-sections of broken-off fragments. All fragments are from pages dated before 1494.

---

[7] Lutz J. Walter, Dipl.-Ing. Restaurator, Tapeten und Bordüren, Alte Weinsteige 110, D-70597 Stuttgart, Telefon: (0711) 7654984.

[8] Deutsche Forschungsgemeinschaft (DFG): „Restaurierung der durch Tintenfraß beschädigten Handschriften des Savigny-Nachlasses. Anwendung der Calciumphytat-Calciumhydrogencarbonat-Behandlung und partieller Stabilisierung in der Praxis", LIS 2 – 557 22 (1) Marburg, INST 1980/1-1, at the Universitätsbibliothek Marburg from 2004 to 2007.

Historic material containing copper pigment ("Kreuterbuch")

Samples were taken from "Kreuterbuch. Darinn unterscheidt Namen und Würckung der Kreutter, Stauden, Hecken ... so inn Deutschen Landen wachsen (etc.)" by Hieronymus Bock, printed 1551 in Strasbourg, now in private property. Several samples from this book were available. The choice for analysis was based upon size of the fragments and state of degradation evaluated by visual analysis. The fragment from page 44 was considered to be most suitable for fluorescence labelling and GPC analysis, because its size allowed for multi sub sampling. Its condition was classified to be rather bad, because both paper and pigment have turned brown and fragments appear to be brittle. For LA-ICP-MS, a fragment from page 46 was chosen that was classified to be in good to average condition because the pigment still appeared green and adjacent paper was not heavily discoloured.

Historic material containing copper pigment ("Kostümbuch")

Samples were taking during a restoration treatment from a renaissance book describing costumes. The book is owned by the Austrian National Library and belongs to the theatre collection (622.191-C rara). The damage caused by the copper pigments found on the paper appeared to differ from visual analysis. However, the two fragments available for analysis in this work are described as heavily damaged when pigment is present and without visually perceivable damage on pure paper. Two of the analyzed compounds were identified as neutral verdigris ($Cu(CH_3COO)_2 \cdot H_2O$) and langite ($Cu_4[SO_4(OH)_6]H_2O$), important amounts of chloride and sulphur lead to the conclusion that most probably pigments in this book are basic copper chlorides and copper sulphates [144]. Further analysis on degradation products found in these samples are described elsewhere [145].

Historic material containing copper pigment ("Tapete")

During a conservation treatment in the 1980s it was possible to obtain some amounts of sample material from a Chinese wall paper originally mounted in the castle of Schlosshof in Austria. The wall paper is now owned by the Museum für Angewandte Kunst MAK in Vienna. Further description of the wall paper can be found in [146].

3.2 Treatments

Sodium hypochlorite on pulp (β-elimination)

The pulp (BKZO) was oxidized by sodium hypochlorite (Fluka) according to following conditions: wet pulp (50 mg till 10 minutes of reaction time and 100 mg for longer oxidations) was weighed. For the reaction, 2 mL of sodium acetate buffer (0.2 M, pH was adjusted to 5 using glacial acetic acid) and 2 mL of fresh sodium hypochlorite (0.1 M) were added to the pulp in 4 mL glass vials with tight caps. The pulp was left for the intended time of oxidation in a shaking water bath at 40°C. The reaction was stopped by washing the pulp with abundance of demineralised water on a suction flask. If labelling did not proceed immediately the oxidized pulp was stored in the dark at -30°C. Pulp hand sheets were made in Lenzing (www.lenzing.at) according to ISO 3688_1999 Pulps - Preparation of laboratory sheets for the measurement of diffuse blue reflectance factor.

TEMPO-oxidation on pulp (influence of derivatization on labelling procedure)

The pulps (cotton linters and BKZO) were oxidized by *2,2',6,6',-tetramethylpiperidine-1-oxyl* radical (TEMPO) obtained from Sigma Aldrich according to following conditions: in a 100 mL beaker, 40 mg of TEMPO were dissolved in 50 mL of water by stirring. After dissolution, the solution was transfered into a 500 mL beaker. First 100 mL of water and then 1.920 g of sodium bromide (Aldrich) were added. A pH-meter was installed. Slowly, 15 mL of sodium hypochlorite (Fluka) were added. First the pH raised, but it quickly starts to decrease. The pH was maintained at about pH 11 by careful addition of 0.4 M sodium hydroxide (Fluka). A sufficient amount of pulp (in this case 2 g of pulp in total were oxidized to provide sufficient amount of sample material) were suspended in this prepared solution. With a small beaker pulp was taken out of the reaction mixture after the desired intervals. The reaction was stopped by washing the amount of pulp withdrawn from the bulk on a suction flask with abundant amounts of ethanol (technical grade) and demineralised water. If labelling did not proceed immediately the oxidized pulp was stored in the dark at -30°C.

Calcium phytate/ calcium hydrogen carbonate treatment on Whatman filter paper (irongall ink corrosion, Kolbe)

To produce even oxidation of irongall ink, model sample paper was first submitted to mild pre-aging (70°C and 50% relative humidity) for three days before further treatment [147]. Half of the samples were treated with phytate, the other half remained as the reference. The treatment with calcium phytate/ calcium hydrogen carbonate was performed using calcium carbonate (Merck), and phytic acid (Mitsui Chemicals) in solution at a pH of 5.9 [112]. The sample paper was immersed in a bath for 20 minutes, dried and coated with 1% gelatine solution using an airbrush. As gelatine surface sizing is added to the dried paper after aqueous phytate treatment, one set of samples was sized with gelatine without any other treatment in order to see if a stabilisation can be detected. The samples with OU ink applied together with gelatine are referred to as "gelatine", and the samples with a phytate treatment performed on OU ink samples, are referred to as "phytate". For description of ink modifications see table 10.

Calcium phytate/ calcium hydrogen carbonate treatment on model rag (irongall ink corrosion)

The model rag with model irongall ink modifications plotted on were divided into two halves. Half of the sample was treated with phytate, the other half remained as the reference. The treatment with calcium phytate/ calcium hydrogen carbonate was performed according to DFG-protocol[9,10], using calcium carbonate (Merck), and phytic acid (Mitsui Chemicals) in solution.

Manual deacidification and antioxidant treatment on model rag (corrosion caused by copper pigment)

Magnesium alcoholate solution and antioxidants were obtained from Conservación de Sustratos Celulósicos S.A., Terrassa, Spain. To test a high concentration of antioxidants, 1.5:1 (w/w) methyl-*p*-hydroxybenzoate and propyl-*p*-hydroxybenzoate were dissolved in 75 ml of 2-propanole mixed with 100 ml treatment solution, the lower concentrated solution contained 1:1 (w/w) methyl-*p*-hydroxybenzoate and propyl-*p*-hydroxybenzoate in 134 ml treatment solution.

Mass deacidification treatment ("T"-samples, oxidized pulps)

The original papers from the 20th century ("T"-samples) and oxidized pulp hand sheets were deacidified according to the CSC booksaver© deacidification process based on carbonated di-n-magnesiumpropylate in n-propanole diluted by fluorinated hydrocarbon. For process details see US patent 6743336, US patent 194902 and US patent 7211221.

Calcium phytate/ calcium hydrogen carbonate treatment on historic papers ("K" samples containing historic irongall ink).

Each piece was cut into half and each half into three different stripes. Half of the samples were treated with phytate, the other half remained as the reference. The treatment with calcium phytate/ calcium hydrogen carbonate was performed according to DFG-protocol[9,10], using calcium carbonate (Merck), and phytic acid (Mitsui Chemicals) in solution.

---

[9] http://www.uni-marburg.de/bis/ueber_uns/projekte/dfgtinte

[10] Deutsche Forschungsgemeinschaft (DFG): „Restaurierung der durch Tintenfraß beschädigten Handschriften des Savigny-Nachlasses. Anwendung der Calciumphytat-Calciumhydrogencarbonat-Behandlung und partieller Stabilisierung in der Praxis", LIS 2 – 557 22 (1) Marburg, INST 1980/1-1, at the Universitätsbibliothek Marburg from 2004 to 2007.

Removal of irongall ink and copper acetate pigment

For the visualization of oxidative damage the visible ink or pigment lines on paper had to be removed before labelling. The removal was performed in an aqueous solution of 1.9 g EDTA III Titriplex (Merck) in 200 mL water (about 0.01 M). To improve the penetration, 50 mL di-ethylenglycol (99%, M = 106.12 g/ mol) were added. The pH was adjusted with 100 µL glacial acetic acid to 4.7. The paper sample with ink or pigment applied on were immersed into this solution in a glass vial with tight cap and left shaking at room temperature till the ink or pigment were removed. If necessary, the Titriplex solution was exchanged several times. In some cases, mild heating up to 60°C was performed to improve removal. Generally the removal was only controlled with the unaided eye, but in case of the copper acetate samples, LA-ICP-MS was performed to control the removal of copper ions on paper (see 3.7 "Additional analytics").

Reduction treatment (sodium borohydride)

NOVO paper was reduced by sodium borohydride ($NaBH_4$) according to following protocol: 25 mg of sample were disintegrated in a mixer, the excess water was removed on a suction flask and the semi-dry pulp was placed in a glass vial with a tight cap. A solution of 2.51 mg $NaBH_4$ solution in 50 mL demineralised water was prepared by stirring. The pH of the solution was adjusted to pH 10 by careful addition of 0.4 M NaOH. Per sample 4 mL of $NaBH_4$ solution was added to the pulp. The vial was left shaking for 24 h at room temperature. The reaction was stopped by thorough washing with water. Before dissolution attempts, the pulp was stored in dark at 4°C.

Reduction treatment by borane tert-butylamine (TBAB)

NOVO paper was reduced by TBAB according to following protocol: 25 mg of sample were disintegrated in a mixer, the excess water was removed on a suction flask and the semi-dry pulp was placed in a glass vial with a tight cap. An aqueous 0.2 M solution of TBAB was prepared by stirring and mild heating. The pH of the solution was assured to be neutral. Per sample 4 mL of TBAB solution was added to the pulp. The vial was left shaking for 24 h at room temperature. The reaction was stopped by addition of few millilitres of 0.1 M HCl and subsequent thorough washing with water. Before dissolution attempts, the pulp was stored in dark at 4°C.

3.3 Aging

Model iron gall ink on Whatman filter paper (Kolbe)

Treated and untreated samples were subjected to accelerated aging to assess the long-term impact of the different treatments. Aging was performed at 90°C with cycling humidity between 35 % and 80 % relative humidity in a Vötsch Heraeus HC 0020 climatic chamber according to previous aging protocols. The paper was subsequently analysed after 0, 3, 6, 9, 12, and 18 days. This allows not only the comparison of absolute values of molecular weight and carbonyl groups, but also a rough kinetic investigation, *i.e.* the comparison of the rate constants when simple rate laws are applicable.

Tidelines on Whatman Filter Paper

Whatman filter paper with tidelines produced by Dr. Anne-Laurence Dupont at CNRS[11] were subjected to accelerated aging for two and five days. The accelerated aging conditions for the papers were chosen according to ASTM D6819-02e2 standard that was slightly modified by using different tube volume and mass of paper, but keeping always the same ratio air volume/mass paper as in the standard method. The aging temperature was 100°C (+/- 2°C), in a dry-heat oven. Septa used are teflon/ silcone and vials are crimp-cap vials. Otherwise,

---

[11] Centre de Recherches sur la Conservation de Documents Graphiques (CNRS), 36 rue Geoffroy-Saint-Hilaire 75005 Paris

all other conditions - including the paper samples equilibration before and after the aging at 23°C / 50% RH - are the same as in the ASTM standard with:

- tidelines samples: aged in 10 mL-glass vials, 278 mg paper
- above and below tideline samples : aged in 49 mL-glass vials, 1,36 g paper

Model rag papers + copper acetate

Degradation caused by copper ions was simulated with accelerated aging at 55°C and fluctuating humidity from 35 – 80 % relative humidity (RH) every 6 hours to enforce water migration within the paper web. This aging treatment was performed on all paper samples analyzed in this study. The comparatively low temperature was chosen because copper acetate pigment is heat sensitive and will turn brown at temperatures above 60°C [148, 149].

Some paper samples were additionally subjected to artificial aging for seven days at 80°C and 65 % RH. Despite of the heat sensitivity this type of accelerated aging was chosen to make results more comparable to other studies.

Model rag papers + ink modifications

To observe the development of irongall ink degradation, the samples with the different model inks have been subjected to accelerated aging. The aging procedure started with one day of pre-aging to build up an even state of oxidation. The accelerated aging procedure has been divided into two phases. The first phase was designed to encourage mobile ink compounds to start migration processes as observed in naturally aged inks. Therefore aging conditions were set on 55°C and cycling humidity from 35% - 85% relative humidity (6h). This is also in accordance with studies performed on copper acetate pigments and allows for comparison between the two sample sets. After the period of cycling humidity, the second phase aging mode was changed to static conditions at 80°C and 65% relative humidity for two days. Samples were drawn after every single aging day.

Historic papers ("K" samples)

Each sample was than cut into 3 stripes (figure 14), which have been submitted to different aging protocols that refer to previous accelerated aging protocols (see model rag + copper acetate and model rag + ink modifications):

- **K1:** 1d pre-aged (70°C, 50 % relative humidity), 7d dynamic aging (55°C, cycling humidity 35 % and 85 % relative humidity)
- **K2:** 1d pre-aged (70°C, 50% relative humidity), 7d dynamic aging (55°C, cycling humidity 35 % and 85 % relative humidity), 7d static aging (80°C and 65 % relative humidity)
- **K3:** non-aged reference.

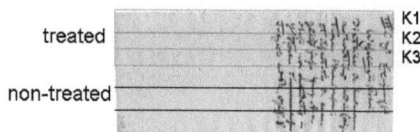

Figure 14. Historic sample and scheme for treatment and aging

Test papers (20th century), "T"-samples

Test papers were aged with and without deacidification treatment in closed vessels at constant humidity (50 % RH) and cycling temperature (20 - 80°C, every six hours) for three weeks. The samples were stored for more than 24 h at the desired relative humidity (50 % RH) to allow equilibration of the paper before sealing them into closed vessels. Aging in closed vessels was found to generate typical degradation products of natural aging [150].

3.4 Sampling

Tidelines on Whatman Filter Paper

Three sub samples were taken from tideline samples (figure 15). One is named "below" and marks the area of the filter paper that was dipped into water, the other is named "above" for the part of the filter paper that is above the tideline region and had no contact with water. The tideline region was cut along the visible brown line that extended roughly about 2 mm. This sub sample is named "tideline". The procedure was the same in lab and under clean room conditions. However, in the clean room polyethylene bottles were used instead of glass beakers and all metal parts were replaced by Teflon.

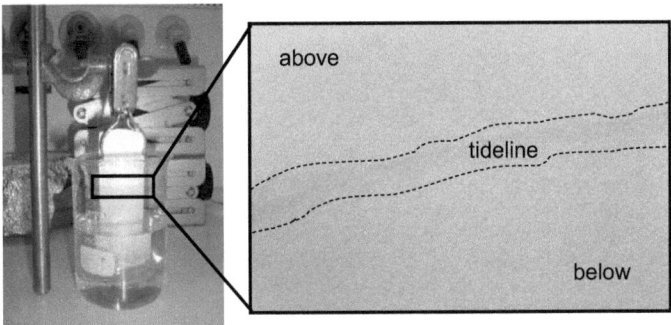

Figure 15. Preparation (left) and sub sampling (right) of tideline samples under lab conditions.

Model rag + ink modifications

Each paper has an ink covered area containing one of the five ink modifications and an edge area. In order to study ink corrosion, one sub sample per sample was always taken from the inked areas and compared to a second sub sample that was taken from the edge area, as far away from the ink application as possible. Using this sampling protocol the influence of the chosen aging procedure on cellulose and on cellulose + ink can be studied.

Model rag paper + copper acetate

The cutting scheme is given in figure 16: So called lines (dashed line rectangles) and blank spaces (black squares) were studied, in addition a mixture of blank spaces and lines, as well as paper from the edge of the paper, which had not been in contact with copper pigment, and therefore should not have been influenced by any migration process.

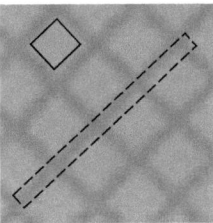

Figure 16. Cutting scheme for paper samples with degradation caused by copper ions. Black dashed line rectangles depict pigment lines, black squares show blank spaces (lines are about 0.5 mm in original).

Historic rag letters

Per sample three sub samples were taken to get some spatial resolution of the damage (figure 17). One sub sample was cut out directly from the inked areas named "ink line", a second one from paper adjacent to ink line named "next to ink line" and a third from paper in at least 1cm distance from inked areas named "edge".

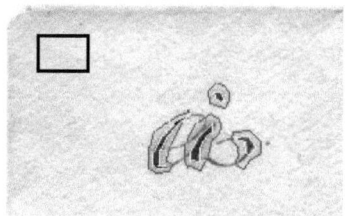

Figure 17. Detail describing sub sample preparation from sample no. 296. The sub sample from the little black square in the upper left part is referred to as "edge". Sub sample "ink line" is described using a white line that surrounds entirely inked areas on the paper sample, sub sample "next to ink line" is described using a red line and covers paper areas directly next to ink application.

"K"-samples (historic rag papers containing irongall ink)

Each stripe, treated or non-treated, aged or non-aged (see figure 14 on page 52), was cut into two sub samples according to figure 18. The "paper" sub sample is pure paper as far as possible away from any ink application, the sub sample "ink" was cut closely along the writing. No attempt was made to cut exclusively the inked lines, because to much damage has to be expected in this case and the success of treatment would be difficult to judge on an extremely degraded material.

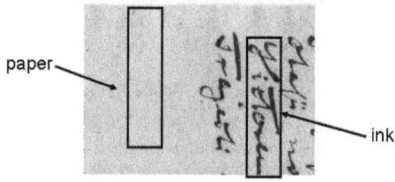

Figure 18. Detail describing sub sample preparation from K-samples.

Test papers (20th century), "T"-samples

A detailed analysis of distribution of surface pH, carbonyl group content and M_w was performed on test paper T3 by dividing it into 30 sub samples (figure 19 left).

The test papers T1, T3, T4 and T18 were divided into nine compartments corresponding to upper row left (1), upper row-middle (2), upper row-right (3), middle row-left (4), middle row-middle (5), middle row-right (6), lower row-left (7), lower row-middle (8) and lower row-right (9) as described in figure 19 right. In this study only results for sub samples (1), (3), (5), (7) and (9) are presented.

1	2	3	4	5	6
7	8	9	10	11	12
13	14	15	16	17	18
19	20	21	22	23	24
25	26	27	28	29	30

Upper row left (1)	Upper row middle (2)	Upper row right (3)
Middle row left (4)	Middle row middle (5)	Middle row right (6)
Lower row left (7)	Lower row middle (8)	Lower row right (9)

Figure 19. Cutting scheme for "T"-samples. Left: Detailed analysis of distribution on test paper T3. Right: Analysis of distribution on test papers T1, T3, T4 and T18.

3.5 GPC and fluorescence labelling

General labelling procedure with CCOA

Carbazole-9-**C**arbonyl-**O**xy-**A**mine (CCOA) labelling of carbonyl groups was performed as described earlier [59-61]. **F**luorenyl-**Di**-**A**zo-**M**ethan (FDAM) labelling of carboxyl groups was performed as described by Bohrn [64]. The general procedure includes the following steps:

- Sampling of pulp or paper (20 - 25 mg)
- Disintegration in water (CCOA)/ 0.1 M hydrochloric acid (FDAM)
- Removal of water (CCOA)/ replacement of water by ethanol (FDAM)
- Labelling for seven days at 40°C in zinc acetate buffer at pH 4 (CCOA)/ in DMAc (FDAM)
- Removal of excess label
- Activation with DMAc (CCOA)
- Dissolution in DMAc/ LiCl (9 %)
- Dilution with DMAc
- Filtration

Micro batch labelling procedure with CCOA

Besides the routine approach with 20 to 25 mg of pulp sample, several micro batches were performed when only 2 – 5 mg of sample material were available, e.g. for multiple sub sampling on historic samples. The sample was disintegrated in water, filtered on a suction flask and labelled in 0.8 mL of zinc acetate buffer and 1.00 mg of CCOA per sample for seven days at 40°C in a shaking water bath. After labelling the excess label is removed by washing with water on a suction flask. The final washing step is performed with ethanol (technical grade) to partly remove water from the sample. Subsequent activation is done in 0.6 mL DMAc overnight on a shaking device. After activation DMAc is sucked of and dissolution is started in 0.5 mL of DMAc/ LiCl (9 % w/v) overnight. When according to visual analysis the solution is complete, 0.75 mL of DMAc are added to the mixture of dissolved cellulose and DMAc/ LiCl. The diluted solution is filtered through a 0.45 µm filtered and measured.

General labelling procedure with dansyl hydrazine

When the labelled samples were not intended for GPC-analysis, but for UV-microscopy, dansyl hydrazine (Fluka) was used to selectively label carbonyl groups in cellulose. The labelling procedure follows the general labelling procedure described for CCOA above. Due to higher molecular weight of dansyl hydrazine, the amount of label needed for an equal amount of pulp had to be recalculated accordingly. For 20 mg of sample to be labelled 4 mL aqueous zinc acetate buffer and 4.04 mg of dansyl hydrazine have been used. Dansyl hydrazine had to be first dissolved in 2 - 5 mL of ethanol (100%) to ease dissolution in the buffer. The dissolution procedure was done at room temperature. The vial for dissolution was covered with alum foil to prevent access of light. The labelling was performed in sealed glass vials at 40°C for seven days in a shaking water bath. After finishing the labelling procedure the sample was washed thoroughly with of water to remove the excess label on a suction flask. The final step of washing included the addition of 50 mL of ethanol (technical grade) to dry the sample and partly replace water. Subsequent drying was performed at room temperature between pulp sheets with light weight to flatten the sample. Ready labelled samples were stored at 4°C in the dark to prevent the label from discolouring.

General derivatization procedure in DMSO and pyridine

The following protocol is based on the optimized protocol developed by Fischer with some minor modifications [151]. The ratio of cellulose/ isocyanate/ DMSO or pyridine respectively was set to 1:100:1000 (w/v/v). In a three neck flask containing a gas inlet, a reflux cooler, an additional sealed opening and placed on a sufficiently powerful magnetic stirrer, a dry cellulose sample (100 mg, corresponding to 1.85 mmol OH groups, activated by freeze-drying from DMAc or water) was suspended in DMSO or pyridine (100 ml) under efficient stirring at room temperature (DMSO) or at 0°C (pyridine) respectively. For 15 minutes, the mixture was stirred in vacuo which initially caused a slight desorption of gas from the suspended sample. After 15 min, nitrogen gas was introduced until atmospheric pressure was reached. The evacuation was repeated for 5 min, and the nitrogen atmosphere was re-established. Phenyl isocyanate (10 ml) or ethyl isocyanate (8 mL) were added over a period of 10 min, and the mixture was heated to 70°C and stirred at this temperature for 2 days to form a clear and transparent cellulose carbanilate solution. After destruction of excess isocyanate by addition of methanol (approx. 2 ml/ml reaction mixture) the solution was precipitated in excess of water and re-dissolved in DMAc/ LiCl (0.9 %) for GPC measurement.

Carbanilation in the presence of MNSO

The reactions followed the above carbanilation procedure in DMSO or pyridine, with the following modifications: only 25 mg of pulp were used, phenyl isocyanate was replaced by 2 ml of ethyl isocyanate, 10 ml of DMSO or pyridine and 2.65 µL of MNSO were used. The carbanilation was carried out according to the above described general protocol. MNSO was added into the carbanilation mixture to replace 1 mol% of the solvent DMSO. After the carbanilation reaction, the mixture was precipitated in water (400 ml) and the precipitated solids washed with n-hexane/chloroform (50 ml, v/v = 1:1) to remove co-precipitated MNSO. The solid remainder was dried in vacuo at room temperature and re-dissolved in DMAc/ LiCl (0.9 %) for GPC measurement.

General set-up of gel permeation chromatography

Measurements used the following components: online degasser, Dionex DG-2410; Kontron 420 pump, pulse damper; autosampler, HP 1100 column oven, Gynkotek STH 585, fluorescence detector *TSP FL2000* (CCOA); multiple-angle laser light scattering (MALLS) detector, *Wyatt Dawn DSP* with argon ion laser (λ_0 = 488 nm); refractive index (RI) detector, *Shodex* RI-71.

The following parameters were used in the GPC measurements: flow, 1.00 ml min^{-1}; columns, four *PL* gel mixedA LS, 20µm, 7.5 x 300 mm; fluorescence detection, λ_{ex} = 290 nm, λ_{em} =340 nm (CCOA), λ_{ex} = 252 nm, λ_{em} =323 nm (FDAM); injection volume, 100 µl; run time, 45 min. DMAc/LiCl (0.9% w/v), filtered through a 0.02 µm filter, was used as mobile phase.

Data evaluation was performed with standard Chromeleon, Astra and GRAMS/32 software.

3.6 Near Infrared Spectroscopy

Near Infrared Spectra were taken on different days using the FTIR spectrometer *MPA™* by Bruker Optics (www.brukeroptics.de) with integration sphere (PbS – detector) for measurement in diffuse reflection mode (15 mm diameter) covering a wave number range from 12 800 cm^{-1} to 3 800 cm^{-1}. Following acquisition parameters were chosen:

- spectral resolution: 8 cm^{-1}
- number of scans for a single sample spectrum: 100
- zero filling factor: 2

For the 110 pulp hand sheet samples pulp was weighed to 2 g portions and processed to lab hand sheets according to ISO 3688 "Preparation of laboratory hand sheets for the measurement of diffuse blue light reflectance factor" by Lenzing AG (www.lenzing.com) without fillers or sizing material. For pulp hand sheets no special caution has been taken for data acquisition sequence, that is, lab hand sheets are assumed to be of sufficient inherent homogeneity to neglect order of data acquisition. Depending on visual pre-selection, different amounts of spectra were incorporated into the modelling procedure of each parameter resulting in different numbers of selected spectra for the different models.

The rag papers were of different ages, covering a range from about the 16th to the 19th century. All samples were stored in dark, but obviously the whole background of these papers could not be reconstructed, therefore further information on storing conditions are lacking. It was attempted to use the same position for wet chemical analysis as that from where the NIR spectra have been taken. For spectroscopic and wet chemical analysis only areas without visually perceivable inhomogeneities have been recorded.

Wet chemical analysis was performed and evaluated according to the general labelling and GPC procedure described above.

All spectra were mean centred, *i.e.* the mean of variable vector has been subtracted from all of its elements [95]. Additionally, derivatives and multiplicative scattering correction were found to be appropriate for the present problem. Except for calibration of rag paper M_w, spectra were processed (smoothed and derived) according to Savitzky and Golay [152] by means of 25-points smoothing filter and a second order polynomial to obtain first and second derivatives, respectively using OPUS software. Multiplicative scattering correction is reported to be a helpful means to account for light scattering as observed with solid samples [95]. First and second derivatives are expected to give best prediction PLS models. Derivatives are resolution enhancement tools that enable the detection of weak peaks that are not visible in the original spectrum. Side effects of derivatives are the loss of original shape. Nevertheless, the ratio of information bearing signal to noise that impacts the quality of calibration remains unaffected [95].

Before and after pre-processing, the spectra have been re-examined for unusual features. Spectra that visually deviated noticeably from the majority have been removed from the calibration and test spectra set.

Spectra evaluation was performed using Bruker *OPUS QUANT 2* for multivariate calibration. This program uses only PLS-1 algorithm for data calibration.

Partial least square regression (PLS-R) models were calculated using Bruker *OPUS QUANT 2* software. For calibration (cross validation) the infrared data sets were regressed against molecular weight (M_w), carbonyl group content, and carboxyl group content, respectively, to find models with high correlations (*i.e.* high correlation coefficient R) and low root mean square error of cross validation (RMSECV). The pulp hand sheets data set was divided into a calibration (cross validation) set and a test set. The cross validated models were then tested through test set validation. The test set was generated automatically using *OPUS QUANT 2* software in the following way. After principal component analysis the number of principal components and the percentage of samples to be selected from the whole data set for the test set have to be defined. Then the selection by the software is done from the aspect of covering the whole concentration range in the best possible way to obtain a robust model.

3.7 Additional analytics

Surface pH

For the determination of surface pH a modified version of TAPPI T 529 om-04 "Surface pH measurement of paper" was chosen. A droplet of demineralised water (20 µL) was applied on the paper surface backed with a soft, non-absorbing support. Measurements were taken with flat surface probe head connected to a SevenEasy™, both by Mettler Toledo (www.mt.com). The duration of the measurement depended on the time for the pH meter to equilibrate. Calibration with standard buffer solutions was performed daily.

CLSM

Images were collected with a spectral Leica SP2 AOBS confocal microscope (Leica Microsystems) equipped with a blue diode laser. The images were coded green.

LA-ICP-MS

For the migration of copper ions analysis was performed by means of direct laser ablation inductively coupled plasma mass spectrometry (LA-ICP-MS). A Nd:YAG 213 nm laser system (ablascope, bioptic, Germany) was coupled to a ICP-MS (Perkin Elmer DRC II, Perkin Elmer, Canada). The ablated material was transported by a 0.7 L min^{-1} He gas flow directly into the ICP-plasma after passing a glass wool filter of approximately 5 mm length which was directly placed into the tubing after the filter. An additional flow of 0.8 L min^{-1} of Ar was added after the filter. This setup corresponds to our standard setup as described in previous work [153]. Analysis was performed on line scans using the laser parameter as shown in table 11. Every measurement cycle started with the acquisition of a blank value which was subtracted from the subsequent transient laser ablation signal. ICP-MS analysis of Cu was performed by analyzing the ^{63}Cu/^{12}C signal. ^{12}C was used as internal standard. The optimized ICP-MS parameters are given in table 12.

Table 11. Laser ablation parameter (ablascope, bioptic, Germany)

Parameter	Value
Wavelength	213 nm
Energy (fluence)	8 J/cm^2
Spot size	50 µm
Scan speed	16 µm / s

Table 12. Optimized ICP-DRC-MS parameter (Elan DRC II, Perkin Elmer, Canada)

Parameter	Value
Nebulizer	PFA
Spray chamber	Cyclon
Nebulizer gas flow	0.8 L min^{-1}
Auxiliary gas flow	1.275 L min^{-1}
Plasma gas flow	15 L min^{-1}
ICP RF power	1200 W

Further paper analysis on samples with wet-dry interfaces, ink lines and copper pigment were performed by means of direct laser ablation inductively coupled plasma mass spectrometry (LA-ICP-MS). A 266 nm laser system (LSX100, Cetac technologies, Omaha, Nebraska) was coupled to an ICP-MS (Perkin Elmer DRC e, Perkin Elmer, Canada). The ablated material was transported by a 1.0 L min^{-1} Argon gas flow directly into the ICP-plasma after passing a glass wool filter of approximately 5 mm length which was directly placed into the tubing after the laser cell.

The LA-ICP-MS scan was performed before and after calcium phytate treatment. Starting on non-inked paper, crossing the ink line and ending on paper without ink again, a line of 4 mm was scanned with a spot size of 50 µm in order to study the distribution of calcium, iron and copper ions.

4 Results and discussion

4.1 Description of degradation in historic papers

The description of historic and model papers helps to characterize the degradation and its spatial distribution. When analyzing a big amount of historic papers some general characteristics can be found. After applying GPC multi-detector set-up some insight can be gained about what condition can be expected in naturally aged papers. Model papers can thereafter be compared to them. Additionally, a micro-destructive method allows for following the distribution of degradation within one single sheet of paper.

General characteristics

When plotting molecular weight distributions of arbitrarily chosen historic rag papers (described on page 48), one common feature is found in most of them: a distinctive shoulder in the region of 5000 to 25000 g/mol (DP 30-150). The determined molecular weight of this shoulder corresponds to hemicelluloses, but also to lower molecular weight degradation products of cellulose. It is difficult to distinguish between hemicelluloses and these shorter cellulose chains. Nevertheless, as the low molecular weight shoulder appears in almost all samples investigated, it is considered to be typical of naturally aged samples (figure 20).

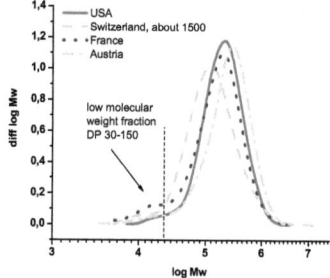

Figure 20. Historic rag papers from different sources and countries. The characteristic low molecular weight shoulder is found in all arbitrarily chosen samples.

The analysis of historic papers with increased damage caused for example by irongall ink corrosion shows that the closer to the ink line the analysed sample was taken, the more important this low molecular weight shoulder gets (figure 21 left). As a growing amount of hemicelluloses is excluded, lower molecular weight cellulose fragments formed after severe hydrolytic and oxidative degradation must account for the observed accumulation. Figure 21 right gives the refractive index signal (mass-proportional) and the carbonyl-proportional fluorescence signal. The region of the above mentioned shoulder shows clearly the high number of oxidized functionalities in the low MW peak.

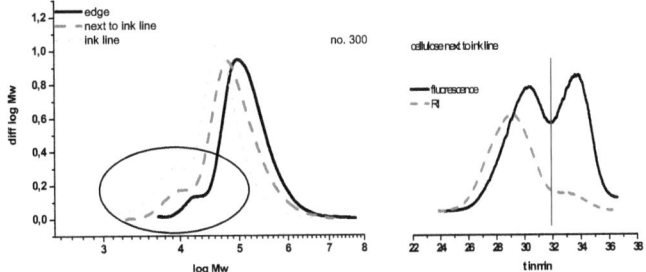

Figure 21. Left: MWD of naturally aged cellulose taken from a historic letter (no. 300, described on page 48) showing the distinctive shoulder in the low MW region that gains more importance when iron gall ink was applied. Right: RI and fluorescence signal for bulk material and low MW shoulder. Most fluorescence is found in the low MW region.

In some of the model papers the increase of this low molecular shoulder that is found in historic papers can be simulated in accelerated aging, especially when oxidative damage occurs due to irongall ink or copper pigments. During accelerated aging of model rag paper (described on page 46) without any applied ink or pigment, the importance of the low molecular weight shoulder is very low (figure 22 left). At the beginning of the simulated aging period, both, paper and paper with copper acetate pigment, have about the same mass percentage of low molecular weight fraction. It is assumed that this fraction rather originates from hemicelluloses that are present in minor amounts in rag paper than from degradation products. Upon aging, the percentage of low molecular weight hemicelluloses in pure paper decreases, probably because some of the molecules found here are further shortened during aging and are not detectable any longer (figure 22 right). Opposite to that, in simulated cellulose degradation caused by copper ions, the mass percentage of low molecular weight fraction increases. Thus, the difference must originate from cellulose degradation products.

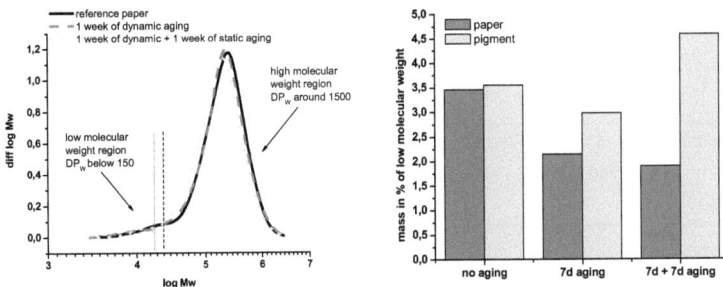

Figure 22. Left: MWD for model rag paper used for the simulation of degradation caused by copper ions in the course of accelerated aging. Right: The percentage of low MW fraction in pure paper is decreased throughout the aging procedure while it is increased when copper acetate pigment is applied.

Especially in Whatman filter paper with unbalanced irongall ink applied on it, this low molecular region gains significant importance (figure 23). As Whatman filter paper consists to almost 100 % of cellulose (see description on page 45 and 46), all products found must originate from cellulose. Thus, low molecular weight cellulose degradation products are accumulated in the low molecular weight shoulder like hemicelluloses.

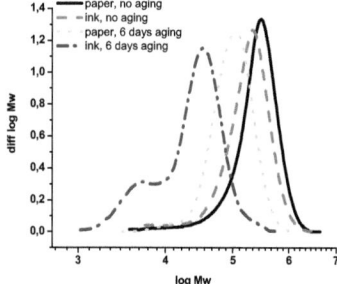

Figure 23. In Whatman filter paper the low molecular weight shoulder only appears in the course of accelerated aging after irongall ink application.

General statistics

During the study of historic rag papers (described on page 48), more than 200 samples have been analyzed. Some general statistics can be applied to get an insight into what parameters in naturally aged papers would typically occur. Next to the low molecular weight shoulder that contains hemicelluloses and degradation products, a historic rag paper is expected to have about 24 µmol/g carbonyl group content and 22 µmol/g carboxyl group content. Its molecular weight is slightly below 200 kg/mol, while surface pH is moderately acidic. Historic rag papers are typically 0.13 mm thick and their brightness is slightly below 60 % reflection for ISO brightness measured at 457 nm in comparison to a barium sulphate white standard.

Table 13. Mean of rag papers for several parameters (different number of n is due to varying amount of analysis of every single parameter)

	Paper thickness [mm]	Carbonyl group content [µmol/g]	Carboxyl group content [µmol/g]	M_w (FDAM-Astra) [kg/mol]	pH surface	Brightness [% reflection]
Mean of n samples	0.13	24.1	21.8	186	5.6	58.5
n (amount of samples available)	266	279	231	232	243	205

Even though not all papers obey to this general summary of cellulose characteristics, a classification would be possible upon the parameters summarized in table 13. Papers having equal amount of oxidized functionalities and molecular weight like described above could be categorized as "average preservation condition". Less oxidized functionalities and higher M_w can be interpreted as "well preserved papers", while more oxidized functionalities and lower M_w correspond to "badly preserved papers". However, this classification needs further refinement.

General observations from the summary table are also reflected in the normally distributed graphs in figure 24. Regarding surface pH, the measured value is strongly related to the most common surface sizing agent used for historic rag papers, gelatine. It is slightly acidic and so is the paper surface. According to figure 24a, it is very unlikely to find a surface pH below 4.5 and above 6.5 in historic rag papers. Only about 13 % of all analyzed papers will have values beyond this most probable range. For both cases, a surface pH below 4.5 or above 7.0, the probability is as small as 5 %.

Paper thickness is linked to its ability to resist mechanical and even chemical attacks [154]. The probability to find papers with less than 0.1 mm paper thickness is below 10 %, and the same is true for papers with more than 0.2 mm thickness (figure 24b).

The findings for surface pH and paper thickness in historical rag papers are explained by the production parameters they have in common, *i.e.* manual production and gelatine sizing, leading to quite uniform parameters. For oxidized functionalities, the same similarities among all rag papers analyzed can be found. Roughly half of the papers investigated have a carbonyl or carboxyl group content close to the calculated mean as can be found in table 13. It is very unlikely to obtain values below 10 µmol/g for both oxidized functionalities. Less than 5 % of the analyzed rag papers have lower carbonyl group content than 10 µmol/g, and less than 1% of the analyzed rag papers have lower carboxyl group content. Values above the calculated mean are more frequently found for carbonyl groups than for carboxyl groups (figures 24c and 24d). As carboxyl group formation is the last possible step in oxidation processes and only aldehyde groups can be further oxidized, this result appears very reasonable and explains why increased amounts of carbonyl groups are more frequently found than carboxyl groups.

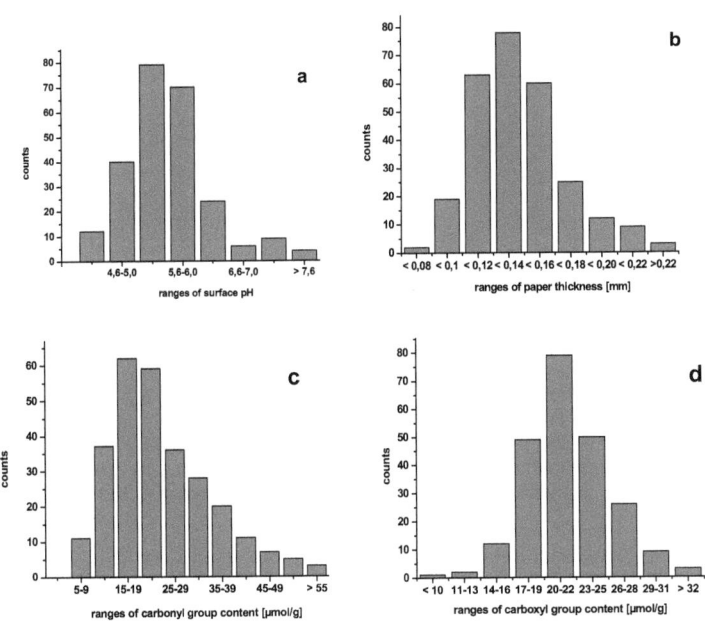

Figure 24. Normally distributed histograms of several rag paper parameters. Parameters are summarized in categories to allow for normal distribution. a. Surface pH. b. Paper thickness. c. Carbonyl group content. d. Carboxyl group content.

Distribution of degradation within one single sheet, "T"-samples

Another way of characterizing historic papers can be used to describe the uniformity of degradation or aging. By dividing a sheet of paper into several sub samples representing the whole sheet, the distribution of surface pH, carbonyl and carboxyl group content and molecular weight can be analyzed. The results are given in colour-codes to make them easy to compare. The papers used for this analysis were obtained from Preservation Academy GmbH Leipzig (see page 48 for description).

A detailed investigation by dividing test paper T3 into 30 sub samples was performed before aging (figure 25). There is a strong correlation of damage in all three parameters in the middle of the upper row: carbonyl group content is increased while M_w and surface pH are decreased in exactly the same sub samples. Surface pH measurement is known to be not as precise as extraction methods. Except from these three sub samples surface pH is homogenously distributed, the range of 4.7 to 5.0 has only low significance. For carbonyl group content and M_w more changes can be detected, because the fluorescence labelling is more sensitive than surface pH. Both parameters can differ considerably within very narrow spatial distribution.

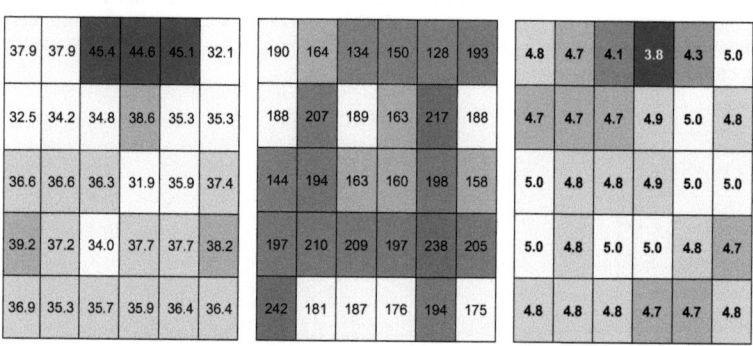

Figure 25. Distribution of carbonyl group content (left), M_w (middle) and surface pH (right) in test paper T3 before aging.

Further investigations were performed on test papers T1, T4 and T18, but they were only divided into nine sub samples (see figure 19 on page 55). Before aging only one sub sample per test paper was measured. As all of these papers had different M_w and pH parameters to start with, the colour code is translated into "good", "average" and "bad" relative to general condition of the respective test paper. The data measured are written inside the single boxes that correspond to the compartment size from which the sub sample was taken. When no sub sample was taken, compartments are filled with stripes extrapolating the neighbouring compartments.

The determination of molecular weight in the tested sheets did not show any regular mapping pattern (figure 26-28 left). Some tendency towards more degradation at the edges of the analyzed papers can be found. Additionally, when assuming a general standard deviation of 5 - 10 % on determination of molecular weight, only test papers T4 (figure 27) and T18 (figure 28) have significantly different M_w values.

Like in the more detailed analysis of test paper T3, surface pH cannot detect many significant differences within the paper sheet. Observed values did not vary much within the investigated paper sheets: the range was maximum 0.4 units. An explanation might be that solubility of alkaline fillers is too low to contribute to surface pH. The surface pH values obtained do not exhibit a generally valid pattern to be interpreted in terms of degradation distribution (figure 26-28 right). Lowest surface pH values are more frequently found in central or lower parts (figure 26 and 28 right). The expectation that due to accumulation of acid degradation products in the middle of a closed book the surface pH of the analyzed papers will be lowest in the central part of the sheet can be confirmed in most papers, but not in all.

Another important conclusion from this type of analysis is that obviously there is no general correlation between surface pH and M_w in the investigated papers. Sometimes sub samples with lowest surface pH correspond to lowest M_w, but this is not a general observation.

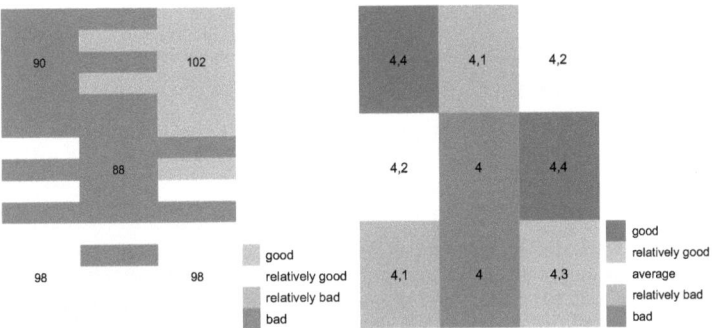

Figure 26. Paper T-1 after accelerated aging. Left: M_w in kg/mol within one sheet of paper. Right: Surface pH within one sheet of paper.

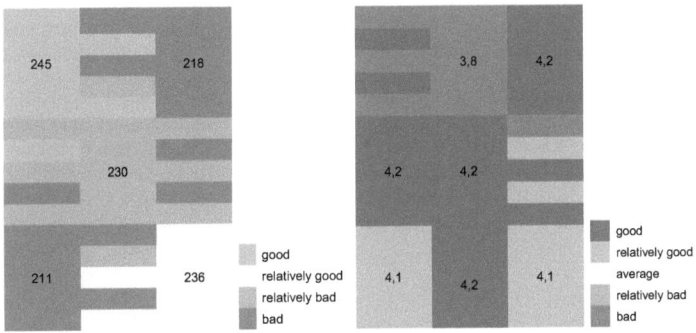

Figure 27. Paper T-4 after accelerated aging. Left: M_w in kg/mol within one sheet of paper. Right: Surface pH within one sheet of paper.

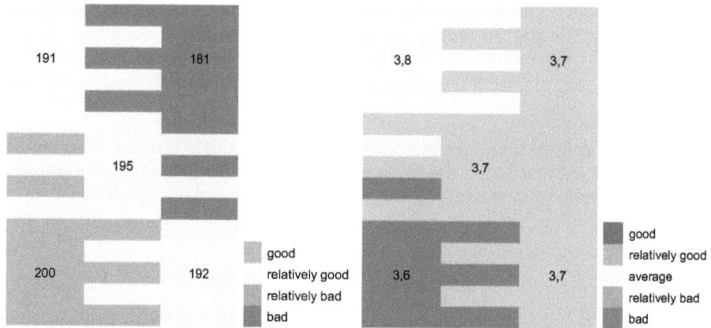

Figure 28. Paper T-18 after accelerated aging. Left: M_w in kg/mol within one sheet of paper. Right: Surface pH within one sheet of paper.

Visualisation of damage

The visualisation of distribution of degradation within one single sheet of paper was approached by using a UV-active label that can be detected with a common UV-lamp at 254 nm. Some attempts have been made to label historic papers with dansyl hydrazine to investigate the distribution of oxidative damage throughout the whole paper sheet. It was found that due to unspecific fluorescence originating from general aging processes that was present in the paper sheet before labelling and that was not removed during a pre-washing step before fluorescence labelling, the interpretation of the amount of specific label was strongly hindered. It is questionable if the required long term labelling (one week) is justified for achieving only few data about oxidation on a relative scale, because no quantification was obtained. The idea of producing a relative scale for semi-quantitative standards was abandoned as it turned out to be difficult to homogeneously label pulp samples of known carbonyl group content for comparison. The question how to perform the labelling on historic originals was not further addressed. The label (dansyl hydrazine) turns into deep blue after exposure to day light and can therefore not be applied on works of art to test their condition. As this approach was intended for in-house use in conservation workshops further efforts do not appear to be justified for the reasons mentioned above.

4.2 Influence of transition metal ions on cellulose and paper

In the following chapters, the results of some studies on model papers, either Whatman filter paper or model rag paper, and historic originals with both, irongall ink and copper containing pigments on them are presented (figure 29).

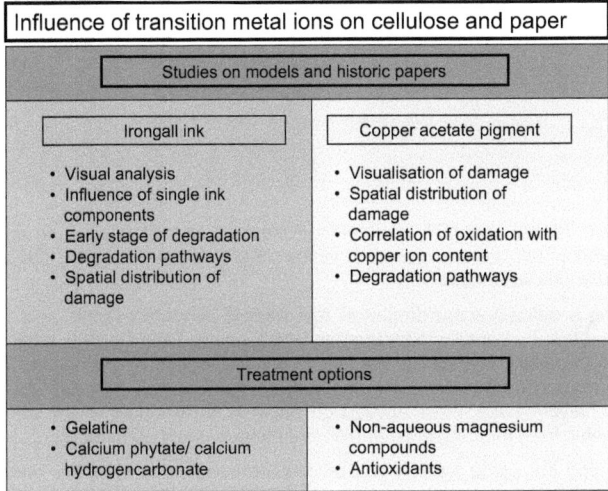

Figure 29. Influence of transition metal ions - overview of topics

4.2.1 Irongall ink corrosion

Next to visual examination, the analysis of single ingredients of irongall ink in comparison with different ink modifications on model rag papers should prove the expected synergistic effects related to irongall ink corrosion. Investigations of irongall ink corrosion have usually covered long periods of accelerated aging e.g. up to several weeks, mainly because available mechanical testing procedures were not sensitive enough to detect subtle changes in early stages of degradation. Contrary to that, fluorescence labelling is sensitive enough to detect these early changes and analyze them for their characteristics. Until now, irongall ink corrosion is assumed to cause oxidative degradation, but a detailed analysis of degradation pathways of sample material as well as historic originals is still missing. The same is true for spatial distribution of damage caused by oxidation.

Visual analysis

For visual analysis the model rag papers described in chapter 3 (described on page no. 46) were used and compared to each other before and after accelerated aging. The model rag paper series included different ink modifications (see table 10 on page 47) and two single ink components, tannic acid and iron(II)sulphate. All of them were prepared using gum Arabic as protective colloid and binding agent.

Visual analysis shows that the colour of all applied modifications on model rag papers changes after accelerated aging. Single ingredients (iron(II)sulphate and tannic acid) experience most obvious colour changes from a rather pale colour before aging into a more intense and dark one after aging. Contrary to that, all inks (balanced ink, unbalanced ink and unbalanced ink containing 7 % of copper sulphate) only change in terms of hues (figure 30).

Typically, these inks turn from a slightly bluish black into a more brownish black. This brownish colour is often found in historic irongall ink drawings and writings.

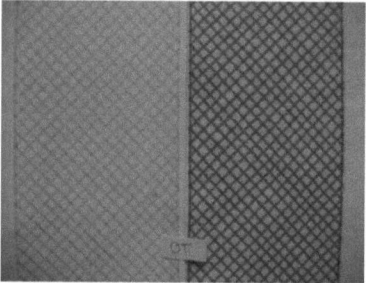 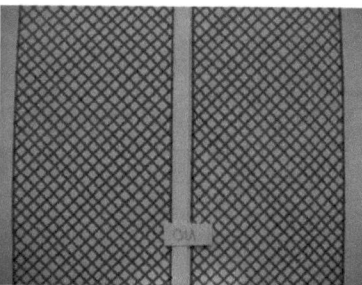

Figure 30. Left: Tannic acid (OT) experienced most visual changes during aging. It turned from a pale yellow into intense greenish yellow. Right: for all inks (OA, OK and OU) only slight visual changes were observed.

When applying a defined water droplet to test wetting behaviour tannic acid leads to most changes from rather hydrophilic to hydrophobic after aging. Tannic acid is additionally able to migrate and form visible deposits according to the distribution of the applied water droplet. Iron sulphate does not lead to pronounced hydrophobic behaviour throughout the observed aging period. Nevertheless, ones applied on paper it is no longer water soluble. Inks are quite hydrophobic from the beginning and do not change much upon aging.

When performing the same type of test on historic irongall ink samples, water absorption takes place preferably along the ink lines. Next to the ink line the paper becomes more hydrophilic upon aging. One explanation for this observation might be degraded gelatine sizing or general degradation of the cellulose. This behaviour was not observed after accelerated aging on model papers, neither on Whatman filter paper nor on the model rag samples.

When removing ink components or model irongall ink respectively, and labelling the paper with dansyl hydrazine, no increased fluorescence was found. Neither former ink lines nor paper in close vicinity of the former lines show any sign of carbonyl group formation.

From these observations it was concluded that unreacted tannic acid in irongall inks remains soluble over quite a long period of accelerated aging, and can therefore contribute to the migration phenomena into the surrounding paper. Dansyl hydrazine labelling suggests that the chosen aging procedure did not induce any strong oxidation.

Single ink components

The same type of model rag papers that according to visual analysis have not undergone any severe damage that might be compared to natural irongall ink corrosion have been analysed for oxidative changes by fluorescence labelling followed by GPC. As a reference, a mean of the paper taken from the edge of the sample material that was not influenced by any applied substance was calculated and the areas affected by an application of ink or ink ingredient contrasted to it.

The carbonyl group content of the paper from the edge area that is not affected from single ink ingredients or ink modifications increased from 3.4 µmol/g to 4.1 µmol/g. This is no significant oxidation of cellulose in the paper. The overall effect on carbonyl group development of inked paper areas was small. No sample exceeded 10 µmol/g carbonyl group content, which is still a comparatively low value. No strong development of carbonyl groups takes place in the paper mean, in iron sulphate (OE) and in tannic acid (OT) samples. Especially in tannic acid-samples almost no oxidation occurs at all. The detected amounts of carbonyl groups are more or less equal to the paper mean. Compared to that the iron sulphate-sample is slightly more oxidized right in the beginning, but no development takes place during aging. All three inks show more oxidation than single ingredients, their carbonyl group content is increased (figure 31).

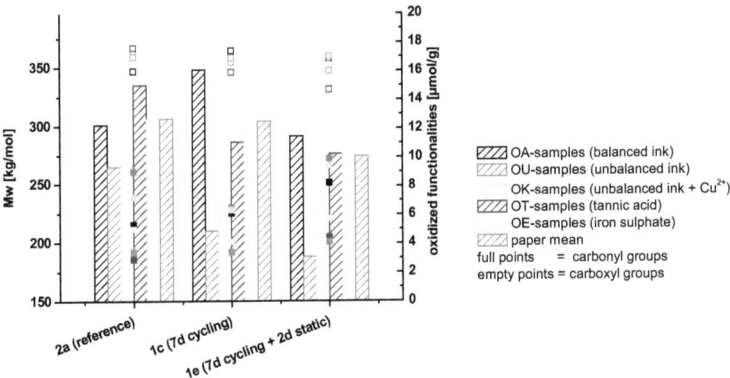

Figure 31. Development of oxidized functionalities and M_w on model rag papers with ink modifications after accelerated aging. While M_w and carbonyl group content slowly change during the aging process, the carboxyl group content is not influenced.

Further oxidation of carbonyl groups leads to carboxyl groups. There is no pronounced carboxyl group development in any of the ink samples. In contrast to that, iron sulphate (OE-samples) experiences some further oxidation to carboxyl groups. This behaviour is neither reflected in their carbonyl group development nor in other iron sulphate containing inks where un-reacted iron ions could have the same effect, *i.e.* promotion of further oxidation. Opposite to that, the amount of carboxyl groups is slightly decreased in tannic acid (OT) samples (figure 31).

Single ingredients (OT and OE samples) and balanced irongall ink (OA samples) do not cause more decrease of molecular weight in the course of aging than the paper mean experiences. In contrast to that, there is a significant decrease in M_w in both unbalanced inks, OU (unbalanced ink) and OK (unbalanced ink with addition of copper ions), the final value is 188 kg/mol respectively 224 kg/mol (figure 31).

Another series of test papers was prepared on Whatman filter paper (described on pages 45 and 46) to study the influence of irongall inks, especially those with unbalanced ratio and addition of copper ions, in more detail with prolonged aging.

The stability of different sample modifications can be compared by calculating the rate of change over a period of time. The rate is derived from the k-value of the slope of a single parameter belonging to one sample over a period of time. It describes the increment. The faster a reaction the steeper is the obtained slope and the higher is the k-value. If an unbalanced ink is applied on Whatman filter paper the total number of carbonyl groups increases as expected. The rate of carbonyl formation is about ten times faster as compared to the reference paper. The degradation is dominated by hydrolytic chain cleavage, the rate of cellulose oxidation increases about five fold over pure cellulose (table 14).

Table 14. Overview of increments for total C=O, oxidized units and REG for Whatman filter paper without ink added as reference paper, unbalanced ink (OU) and unbalanced ink with copper ions added (OK).

	Reference [μmol/g]/d	Unbalanced ink (OU) [μmol/g]/d	Unbalanced ink + Cu^{2+} (OK) [μmol/g]/d
Total C=O	0.38	3.77	4.65
Oxidized units	0.17	0.86	1.57
REG[1]	0.21	2.91	2.81

[1] REG: Reducing end groups calculated from M_n (please note: this value assumes that reducing ends are not further oxidized and that determination of M_n is accurate)

The MWD shows first signs of accumulation of lower MW degradation products after 12 days of aging. The course of M_w and M_n over time (figure 32) does not follow a simple rate law anymore. The overall rate of degradation is much faster and reaches M_w values of 37 kg/mol after 12 days of aging which corresponds to a DP around 200.

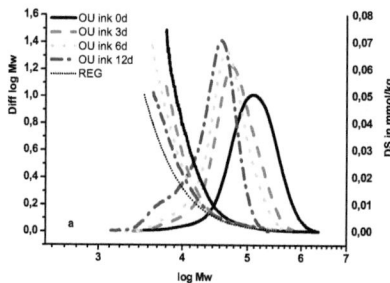

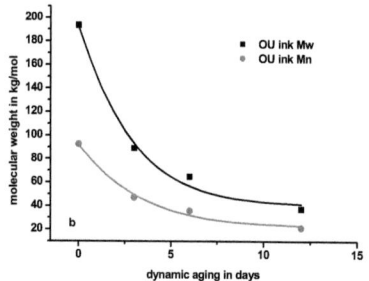
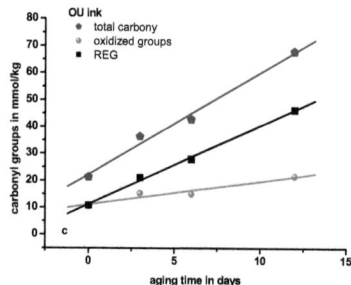

Figure 32. Unbalanced ink on Whatman filter paper. (a): Molecular weight distribution (MWD) depending on the time of accelerated aging. (b): Course of M_w and M_n. (c): (pseudo)-zero order of carbonyl group formation: reducing end groups (REG) are indicative for hydrolytic processes, oxidized units are calculated from total C=O minus REGs and illustrate reactions involving oxidation.

The addition of copper ions should even better simulate the natural conditions as they are commonly found in inks on historic documents [154]. Degradation caused by copper ions in general shows a higher potential for oxidative processes, hence it was of interest if this could be simulated also in this model series with irongall inks. From table 14 it is obvious that the rate for the formation of carbonyl groups increases. However, in figure 33c it can be clearly shown, that the addition of copper ions caused predominantly the oxidation of cellulose without a significant influence on the hydrolytic degradation. The final MWD is in the same order of magnitude as pure unbalanced irongall ink. However, the course of degradation is first slowed down and subsequently accelerated (figure 33b) and indicates a more complex reaction kinetic.

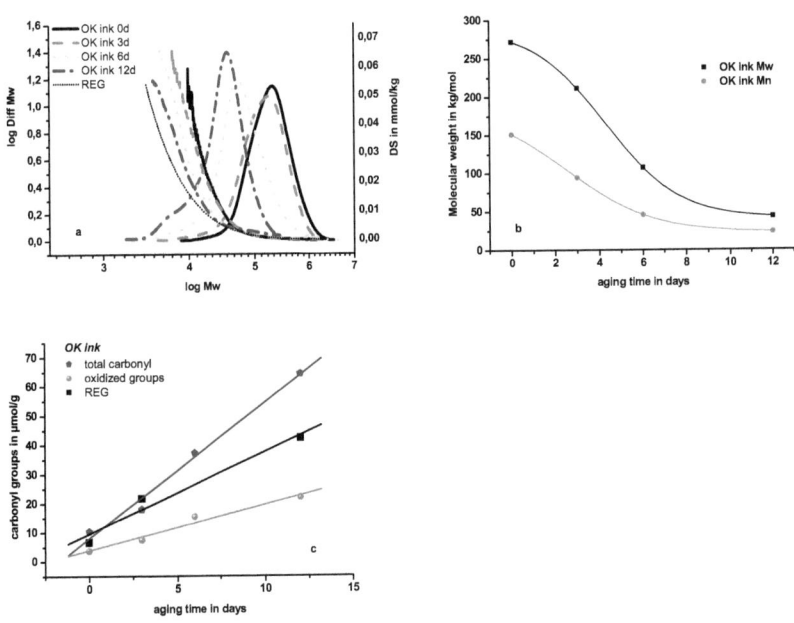

Figure 33. Unbalanced ink with the addition of Cu^{2+} on Whatman filter paper. (a): Molecular weight distribution (MWD) depending on the time of accelerated aging. (b): Course of M_w and M_n. (c): (pseudo)-zero order of carbonyl group formation: reducing end groups (REG) are indicative of hydrolytic processes, oxidized units are calculated from total C=O minus REGs and illustrate oxidative reactions.

Early stage of degradation

The model rag papers used in the previous two sections have been aged for 7d + 2d, while every 24 hours samples have been drawn to gain insight into the early stage of degradation. A closer insight on aging kinetics shows that during the first seven days of aging no significant oxidation takes place in all samples. Most increase in oxidation for the observed period of aging occurred during static aging. Within the relatively short period of two additional aging days, both inks, OU (unbalanced ink) and OK (unbalanced ink with additional copper ions), have almost doubled their carbonyl group content. In table 15 the kinetic results for carbonyl group development are summarized. For the paper mean the k-value of static aging is 0.5 [µmol/g]/d, tannic acid (OT) samples have a similar value, iron sulphate (OE) samples have a slightly stronger increase in carbonyl groups. OK-ink and OU-ink cannot be treated according to the same rate law, the growth in carbonyl group content follows an exponential order.

Table 15. Summary of k-values for assuming a (pseudo)-zero rate law of all ink modifications plus paper mean. During cycling aging no significant increase or decrease can be observed. After static aging, especially inks exhibit an increase in carbonyl group content, while OE and OT have a comparable value like the paper mean.

	OA: Balanced ink [µmol/g]/d	OU: Unbalanced ink [µmol/g]/d	OK: Copper sulphate ink [µmol/g]/d	OT: Tannic acid [µmol/g]/d	OE: Iron sulphate [µmol/g]/d	Paper mean [µmol/g]/d
Cycling aging (2c-1c)	0.1	-0.1	0.0	0.1	-0.1	0.0
Static aging (1c-1e)	1.1	n.d.	n.d.	0.5	0.8	0.5

When studying the kinetics of molecular weight decrease during cycling aging, a slight, but constant decrease in M_w of unbalanced ink modifications can be observed from the first day of aging on (table 16). Cycling aging obviously influences the molecular weight in a more pronounced way than compared to oxidation, i.e. hydrolysis slightly dominates the first period of aging for unbalanced inks. Static aging increases the rate of chain scission for all modifications, but most strongly for OA-ink and OE-samples, followed by the paper mean that is used as a reference material. Interestingly, excess of metal ions in OK- and OU-inks caused less chain scissions than observed in paper mean. However, an excess of iron ions from iron sulphate as a single ingredient (OE) gave a considerable hydrolytic degradation, obviously due to a limited buffering capacity of gelatine or cellulose.

During cycling humidity aging the average molecular weight of model paper without applied substance on it the original value is largely maintained, only after static aging a decrease from 306 kg/mol to 274 kg/mol takes place.

Table 16. Summary of rate constants k for all ink modifications and paper mean for decrease of molecular weight.

	OA: Balanced ink [kg/mol]/d	OU: Unbalanced ink [kg/mol]/d	OK: Copper sulphate ink [kg/mol]/d	OT: Tannic acid [kg/mol]/d	OE: Iron sulphate [kg/mol]/d	Paper mean [kg/mol]/d
Cycling aging (2c-1c)	-1.9	-9.2	-5.6	-2.9	0.9	-1.9
Static aging (1c-1e)	-28.1	-11.0	-8.2	n.d.	-45.2	-14.9

An important observation is an instant oxidation and chain scission occurring with inks directly after its application on the model paper (table 17). Especially with unbalanced ink (OU-ink) and unbalanced ink containing copper (OK-ink) this effect is very pronounced, leading to an increase of carbonyl groups by a factor of 3 without any accelerated aging. The same observations have been made before on Whatman filter paper immersed in aqueous solution of iron(III)chloride and copper(II)chloride by other researchers [155]. Paper without any applied substance and single ingredients do not show this type of degradation. Therefore, natural aging due to the time that passes between sample preparation and measurement can be excluded.

Table 17. Comparison of freshly prepared sample material after and without application of different ink modification, no accelerated aging.

	OA: Balanced ink	OU: Un-balanced ink	OK: Copper sulphate ink	OT: Tannic acid	OE: Iron sulphate	Paper mean
Carbonyl group content [µmol/g]	5.3	9.0	7.2	2.9	5.0	3.4
Molecular weight [kg/ mol]	301	265	267	335	306	306

After ink application, "weak spots" within the cellulose backbone must have undergone an instant degradation [156]. The fact that some degradation takes place immediately after application of unbalanced, corrosive inks while there are no pronounced changes during a quite long period of accelerated aging using cycling humidity can be interpreted as a deterioration process that is divided in at least two phases: a fast initial reaction and a slower induction phase. It was found that the loss in M_w is not solely responsible for the increase in carbonyl groups, also additional oxidation occurs. In order to study where this immediate oxidation takes place, the fluorescence and the RI-signal were plotted against retention time (figure 34).

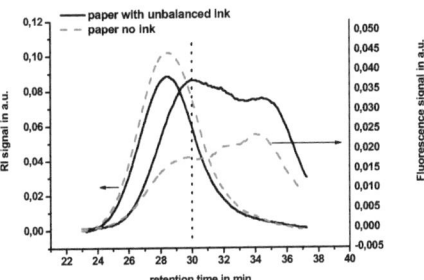

Figure 34. RI and fluorescence signal (corresponding to carbonyl groups) for paper with and without two unbalanced ink modifications; the fluorescence signal is increased at retention times around 30 min, corresponding to an M_w of 200 kg/mol after ink application (see dotted line).

A bimodal function was obtained for inked paper and a trimodal function for paper only. The fluorescence signal detects that oxidation mainly occurs in higher molecular weight regions of about 200 kg/mol for inked paper samples (figure 34, dotted line). This phenomenon was observed for both unbalanced inks. In the course of accelerated aging, the bimodal fluorescence function shifts more and more to an equal distribution between higher and lower molecular weight regions, thus oxidation in lower molecular weight regions gains more importance.

The occurrence of oxidation is furthermore strongly linked to the temperature. As soon as the temperature was increased in the second aging phase, the oxidation proceeded faster, at least for unbalanced ink and unbalanced ink containing copper ions. The amount of carbonyl groups then decreases again after pre-aging.

Degradation pathways and spatial distribution of damage

In historic samples the spatial distribution of ink corrosion was investigated by taking sub samples (see figure 17 on page 54). It was found that corrosion can be limited quite strictly to the inked areas, but just as well stretch over distances about 1 mm away from ink application (figure 35).

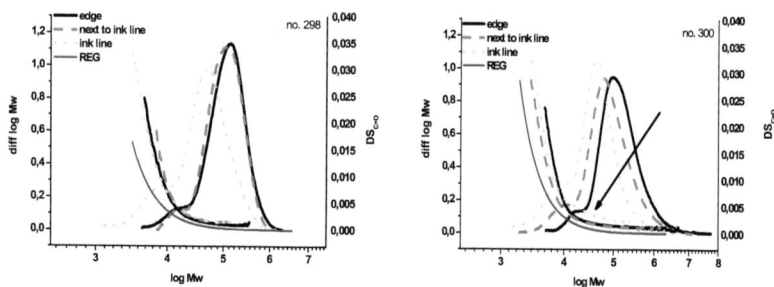

Figure 35. Left: Limitation of deterioration to the inked line in samples no. 298. Right: Damage of ink application on sample no. 300 also affects areas that are adjacent to ink application but not directly covered by ink. The black arrow points at oxidation even in higher MW regions revealed by the $DS_{C=O}$-plot.

Another conclusion was that ink corrosion may be driven by hydrolytic processes, but just as well by oxidation. Sample no. 298 is slightly oxidized in all MW regions, easily detectable when comparing $DS_{C=O}$ plots of all sub samples to DS_{REG}-plot. Considering the $DS_{C=O}$ plot of "next to ink line" and "ink line" that runs quite parallel to the DS_{REG}-plot a possible conclusion is a rather hydrolytic driven influence on the cellulose adjacent to "ink line". The plot of reducing end groups (REG) is estimated from the M_n-data of the MWD. This estimation assumes that all reducing end groups are present as such and not further oxidized, e.g. to the corresponding acid, and that the absolute value of M_n is correctly determined. Next to an overall oxidation of the sample no. 300, "ink line" shows a clear presence of oxidized functionalities. This oxidation obviously affects all molecular weight regions, even higher molecular weight areas.

4.2.2 Degradation caused by copper containing pigments

Model rag papers with simulated degradation caused by copper acetate pigment (see page 46 – 48) showed changed penetration properties of water after accelerated aging. After applying a droplet of water a transition from homogeneously wettable material into paper with hydrophilic and hydrophobic areas occurred. This was interpreted as successful simulation of degradation caused by copper ions [157]. In the following chapters on degradation caused by copper containing pigments this phenomenon was investigated more in detail in order to visualize, characterize and quantify the degradation, define its spatial distribution and link it to the metal ion content.

Visualisation of damage

A detailed analysis revealed that hydrophobic areas in the model rag paper occur directly next to the pigment layer, while blank areas between the pigment lines and areas apart from pigment application stay hydrophilic. Labelling of the damage-aged sample with dansyl hydrazine, a label selective to carbonyl groups, gave further insight into the distribution of carbonyl groups within the damaged paper sample. Before the labelling was performed, the copper acetate pigment had to be removed in an aqueous solution of EDTA. The removal of copper acetate was controlled by LA-ICP-MS. Areas with high fluorescence intensity correspond to areas of high hydrophobicity (figure 36).

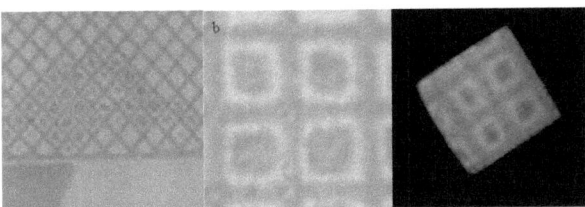

Figure 36. Different wetting behaviour was linked to different degrees of oxidation by group selective fluorescence labelling. The areas that could not be wetted after applying a droplet of water exhibited stronger fluorescence due to increased amount of carbonyl groups detected by selective fluorescence labelling.

Spatial distribution of damage

Simulated degradation induced by copper ions caused pronounced oxidation of the cellulose. The carbonyl group content within in the blank spaces next to the printed test areas was increased to more than three times of the original value (11 µmol/g to 38 µmol/g), while the cellulose at the edge – being remote from pigmented areas – and even below the printed on areas oxidized significantly less (5 µmol/g to 9 µmol/g and 18 µmol/g to 31 µmol/g). CCOA labelling detected differences between line and blank space (figure 37). Interestingly, the values in the blank areas (38 µmol/g) exceeded those from under the pigment lines (31 µmol/g), whereas prior to aging the situation was opposite, 11 µmol/g in the blank space vs. 14 µmol/g under the ink lines. This differentiation was not achieved by determination of molecular weight.

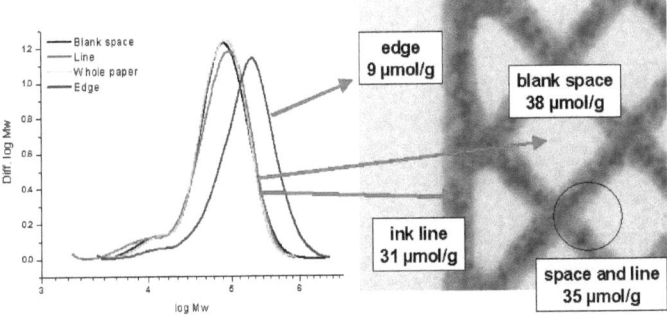

Figure 37. It was possible to quantify the observed different level of oxidation in samples with degradation induced by copper ions, while no significant differences in MWD could be detected (for cutting scheme see figure 17 on page 53).

Table 18 summarizes the data for the model papers obtained according to the CCOA-method. Changes in carboxyl group content are not as pronounced as for carbonyl group content after seven days of aging. Still they show the same trend. Without pigment contact there is least carboxyl group formation. Both sub samples, with direct pigment contact and adjacent to pigment, have increased carboxyl group content, but the sub sample "adjacent to pigment" shows even stronger oxidation than "direct contact".

Table 18. Summary of CCOA data, model paper samples after 7 days of aging

Sample	M_w [kg/mol]	M_n [kg/mol]	M_z [kg/mol]	PDI[1] M_w/M_n	REG[2] [µmol/g]	C=O [µmol/g]	COOH [µmol/g]
No contact with pigment	251	101	501	2.49	9.9	9.5	15.1
Direct contact with pigment	105	47	108	2.24	21.3	31.3	18.4
Adjacent to pigment line	114	61	215	1.89	16.5	37.8	20.1

[1] PDI: Polydispersity index: M_w / M_n
[2] REG: Reducing end groups calculated from M_n (please note: this value assumes that reducing ends are not further oxidized and that determination of M_n is accurate)

After the accelerated aging period the molecular weight distribution shows a general degradation process very clearly: the distributions from material in contact with the copper pigment showed a drastic shift to lower molecular weight indicating chain scissions. High-MW regions and low-MW material were affected likewise. The carbonyl profiles were not identical anymore, but showed a clear vertical shift, typical of oxidative carbonyl group introduction. This offset of the DS plot (figure 38, left) clearly proved the oxidation of the bulk material where copper ions were present. However, in contrast to the CCOA method, the MWD curves cannot distinguish between different locations within the printed areas.

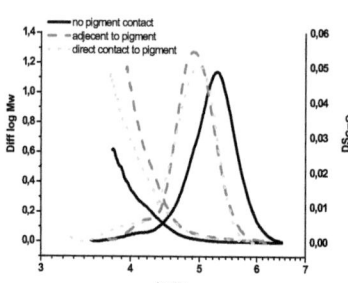 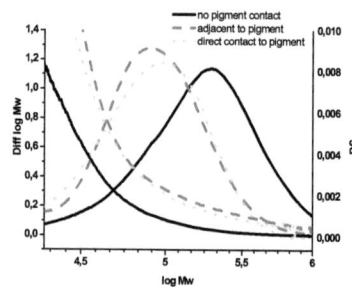

Figure 38. Left: Molecular weight distribution and degree of substitution ($DS_{C=O}$) plots for carbonyl groups of aged model paper (7 days). Right: Enlargement of high MW region of the molecular weight distribution and degree of substitution ($DS_{C=O}$) plots.

The intimate contact with the pigment was not necessary to induce oxidative processes. Overall cellulose oxidation was even higher in areas adjacent to the pigment layer than in regions directly below the pigment. This result emphasizes the role of copper ion migration in degradation caused by copper ions.

Correlation of oxidation with copper ion content

It was demonstrated by LA-ICP-MS that copper ions diffused into neighbouring paper areas upon accelerated aging. This explains the degradation detected in those areas. Without further aging a slight migration of copper ions into the surroundings of a pigment line can be detected (figure 39 left). After seven days of accelerated aging, LA-ICP-MS detects a significant migration of copper ions into the surroundings where there is no protection from binding media (figure 39 right). Significantly higher carbonyl group content in the blank spaces next to copper lines correlates with these findings as transition metal ions will induce catalytic oxidation of cellulose molecules. The shape of the migration curve indicates a diffusion process of copper ions after aging for seven days.

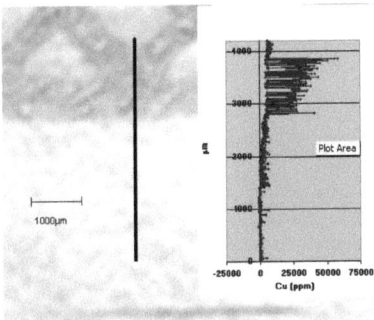 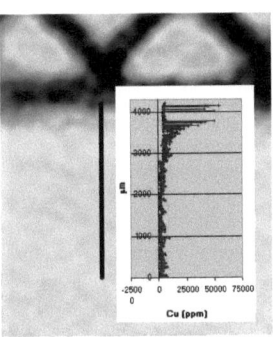

Figure 39. Model rag papers with simulated degradation induced by copper ions were scanned by LA-ICP-MS to determine the quantity of copper ions. Left: Already after simulation of degradation caused by copper ions a significant amount of copper ions migrated into the surrounding paper web. Right: After seven days of accelerated aging the scan from the pigment line into adjacent paper the amount of migrated copper ions even increased.

However, it remains unresolved why degradation in these areas was even more pronounced than directly underneath the pigment. One explanation might be that the copper pigment within the lines will be fixed to a certain extent by protein containing binding media like gelatine or glue, which can have a slightly protecting influence. This could explain why there is less oxidative damage after seven days of accelerated aging on cellulose directly below pigment line.

Degradation pathways

Some fragments of historical paper with copper pigment applied on it were obtained from former Burger Library Bern, Switzerland (see page 48). These fragments originated from different pages of codex 801. One fragment contained a green copper pigment (f. 169), another fragment an additional red pigment (f. 60), and one piece was without pigment cover (f. 137). The red pigment was identified as minium (red lead pigment) and cinnabar (red mercuric sulphide), which are insoluble substances being highly unlikely to be involved in any hydrolytic process or in cellulose degradation. The originally used green copper pigment could not be identified as the degradation caused by copper ions was too advanced [158]. A peculiar feature of these historic samples was the lack of any discoloration of the paper as usually found with degradation caused by copper ions: neither the cellulose nor the green pigments depict any yellowing or showed the presence of chromophores. Also the cross section of the paper was still white. The lack of discoloration was sharply contrasted by the deteriorated mechanical properties, with small pieces falling completely loose.

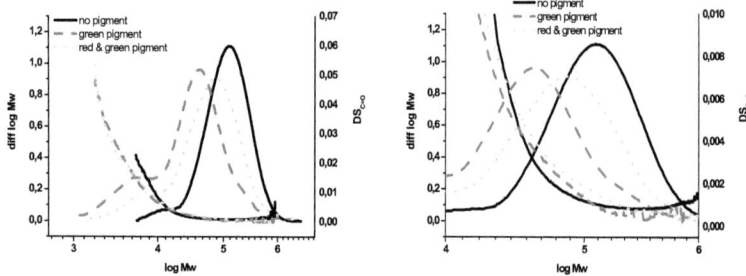

Figure 40. Left: Molecular weight distribution and degree of substitution ($DS_{C=O}$) plots for carbonyl groups of original paper. Right: Enlargement of high MW region of the molecular weight distribution and degree of substitution ($DS_{C=O}$) plots.

Table 19 summarizes the data obtained by the CCOA method applied to the three historical paper samples. The M_w of the pigment-free historic sample represents the action of natural aging without the additional effects of pigment coverage and was found to be comparable to other papers of this age. The degradation of the cellulose fragment containing both green and red pigments was evident, but did not reach the extent as in the fragment with green pigment only. A drastic decrease of the M_w of the cellulose was observed for the fragment with the green pigment on it. In case of the historical material the calculated reducing end groups ranged quite close to the measured carbonyl data. This proved that the cellulose degradation was largely due to hydrolytic influence and only insignificantly due to oxidative impact. Carbonyl profiles of all three samples were similar within the error limits. Neither in the bulk nor in the high MW part significant oxidation effects were found (figure 40 right). The white paper appearance, also in the cross-section, sustains this fact, since oxidative modifications of cellulose usually translate into severe discoloration. However, the pH of the paper was between 6.1 and 6.5. The mechanism of this type of degradation hence remains to be elucidated.

Table 19. Summary of CCOA-data, historical paper samples.

Sample	M_w [kg/mol]	M_n [kg/mol]	M_z [kg/mol]	DP_w	PDI[1]	REG[2] theor.	C=O in [µmol/g]
Cod. 801 green pigment	67	15	289	411	4.57	69.1	68.7
Cod. 801 green and red pigment	113	28	374	697	4.03	35.6	45.2
Cod. 801 no pigment	169	79	319	1039	2.14	12.7	18.2

[1] PDI: Polydispersity index: M_w / M_n
[2] REG: Reducing end groups calculated from M_n (please note: this value assumes that reducing ends are not further oxidized and that determination of M_n is accurate)

Further historic examples containing copper pigments ("Kostümbuch", "Kreuterbuch" and "Tapete") were obtained for analysis (table 20). The degradation pathway in these samples was oxidative as commonly expected. In most cases analyzed, the "pigment" sub sample was far more degraded and exhibited more oxidation than the "paper" sub sample (figure

41). An unexpected feature was the development of a trimodal MWD. This contrasts sharply with the findings in chapter 4.1 where a bimodal distribution (cellulose bulk material in high MW region and degradation products plus hemicelluloses in the low MW region) was found to be typical for historic rag papers.

Table 20. Summary of CCOA-data, Kreuterbuch, Kostümbuch and Tapete samples.

Sample	M_w [kg/mol]	M_n [kg/mol]	M_z [kg/mol]	PDI[1]	REG[2] theor.	C=O in [µmol/g]
Kostüm, paper	194	37	544	5.28	27.2	37.7
Kostüm, pigment, page 113	227	18	1113	12.90	56.9	73.6
Kostüm, paper, page 87	177	47	967	3.73	21.2	52.6
Kostüm, pigment, page 87	197	15	886	13.02	66.1	77.5
Kreuter, paper, page 44	194	28	699	6.82	35.1	66.6
Kreuter, border, page 44	292	25	1643	11.70	40.1	63.2
Kreuter, pigment, page 44	331	45	794	7.33	22.2	25.3
Tapete, white	466	95	1244	4.92	10.6	34.3
Tapete, bright green	285	24	1507	11.82	41.5	71.5
Tapete, green	461	33	1573	14.07	30.5	95.3
Tapete, blue	118	28	324	4.27	36.2	95.3

[1] PDI: Polydispersity index: M_w / M_n
[2] REG: Reducing end groups calculated from M_n (please note: this value assumes that reducing ends are not further oxidized and that determination of M_n is accurate)

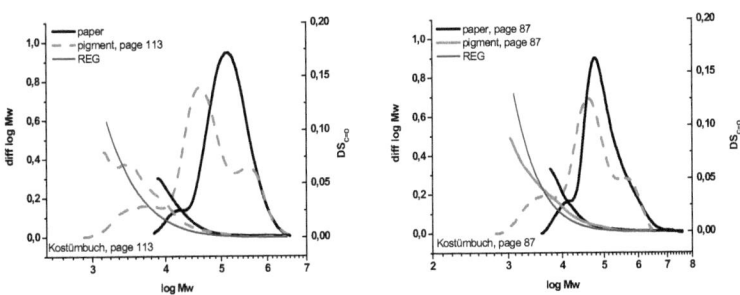

Figure 41. Molecular weight distribution and $DS_{C=O}$-plots of page 113 (left), and page 87 (right) from "Kostümbuch". Sub sample "pigment" shows a trimodal distribution and is far more oxidized.

A sample taken from "Kreuterbuch" page 44 was of sufficient size for multi sub sampling. Therefore it was possible to compare the "paper" sub sample to the "pigment" and "border" sub samples. The "border" sub sample was taken from the paper adjacent to pigment application. Contrary to observations made in all irongall ink samples and most copper pigment samples, the "paper" sub sample suffered from most degradation, while at least parts of the "pigment" sub sample are better preserved (figure 42). Interestingly, most obvious changes in MWD occur in the high MW region, the low MW shoulder is very similar in all three sub samples. Again, the "pigment" sub sample developed a trimodal MWD like already found in "Kostümbuch" samples. When comparing the $DS_{C=O}$-plots of all three sub samples to the DS_{REG}-plot, most oxidative damage is found in the „paper" sub sample. In the "pigment" sub sample, oxidation only gains importance within the low MW shoulder.

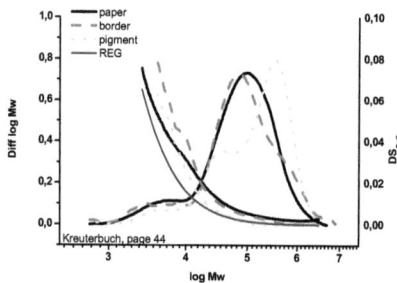

Figure 42. Molecular weight distribution and $DS_{C=O}$-plots of the historic sample "Kreuterbuch" page 44.

Another interesting observation was made when analyzing fragments of a wall paper ("Tapete") that has been printed on with various pigments. Historic wall papers were traditionally imported from the Far East and painted on a complex layer of different papers from mulberry fibres [159]. When comparing the MWD of figure 43 derived from European rag with the MWD of the Asian wall paper sample the difference between the two types of fibres can easily be detected: the MWD of the wall paper sample is shifted to higher MW. The best preserved sample is the one having a white pigment on it (figure 43). That is not surprising as it was most probably chalk.

Worst results in terms of MW were obtained from the blue pigment. This was not expected, as most commonly blue pigments are not reported to be aggressive towards cellulose. The blue pigment sample is also oxidized as can be seen when comparing the $DS_{C=O}$-plot to the DS_{REG}-plot. Most oxidation took place in the lower MW peak, while in the higher MW peak both plots, $DS_{C=O}$ and DS_{REG} do not differ much from each other.

There were two samples of green pigments available on the wall paper. The bright green pigment has a common bimodal distribution and some oxidation, but compared to the white pigment samples the MWD has shifted considerably towards lower MW. In contrast to that, the green pigment shows a trimodal MWD and pronounced oxidation. The influence of the pigment has additionally changed the form of the MWD in a strong way. Obviously the pigment or the mixture in which it has been applied seems to play an important role in how cellulose will be degraded in historic sample material.

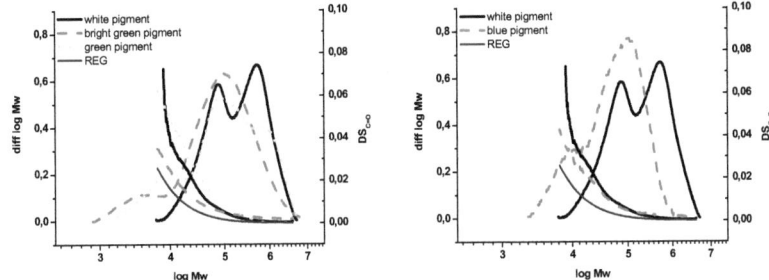

Figure 43. Molecular weight distribution and $DS_{C=O}$-plots of fragments from a wall paper containing different pigments. The sample with the white pigment gives the typical bimodal MWD while no low molecular weight fraction is contained. Left: White pigment and two different shades of green pigment. Right: White pigment and blue pigment.

Most of the analyzed samples containing copper pigments show the tendency to form trimodal molecular weight distributions, an observation that is generally not found in cellulose analysis. Especially when oxidation plays an important role this unusual distribution can be observed. In most cellulose samples the MWD is bimodal consisting of the bulk material (cellulose) and a lower molecular shoulder containing hemicellulose and some degradation products of cellulose. Some samples, like cotton linters even have a monomodal distribution, because they mainly contain cellulose. In the trimodal distribution some part of the cellulose appears to be in a similar condition like the reference material it is compared with, while the bulk part shifted towards a more degraded condition, and additionally the typical low molecular weight shoulder is formed. The origin of the trimodal MWD is not known so far.

One hypothesis for this unusual MWD in copper pigment containing samples is cross linking. Cross linking might occur between several cellulose molecules or within one cellulose molecule. In both cases, aldehyde groups formed upon oxidation can be the reason for cross linking reactions. Copper ions will indeed lead to an increased amount of oxidized functionalities available for this type of reaction. As the hydrodynamic radius of such cross linked molecules will change in dissolution, laser light scattering can be used to detect this phenomenon. Nevertheless, especially in pigment containing samples no indications like sudden bends or curvature can be found (figure 44).

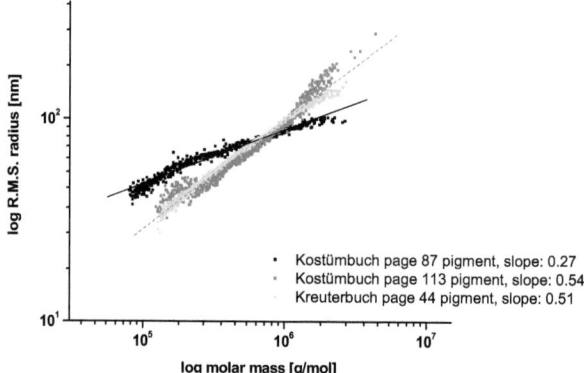

Figure 44. When the RMS radius is plotted versus molar mass the state of dissolution of a molecule can be described. The slope of Kostümbuch page 87 is less steep than for the two other samples indicating a more compact sample. Generally no change in slope that might explain a trimodal MWD can be found even though all three samples lead to such a distribution (compare to figures 41 and 42).

Another hypothesis to explain the formation of a trimodal MWD is the preservation of a part of the cellulose molecules in a condition comparable to reference material. Thus, copper pigments might cause a rather selective degradation of either pre-damaged or more accessible cellulose molecules.

Having analyzed several different historic samples that suffered from degradation caused by copper ions it can be concluded that like for irongall ink corrosion both degradation mechanisms, hydrolysis and oxidation, can be found. In most of the analyzed samples though, oxidation is the predominant pathway.

Correlation of oxidation with copper ion content

As for irongall ink corrosion samples also one original sample with degradation caused by copper ions was investigated by LA-ICP-MS to find out about the spatial distribution of copper ions within the paper. First of all an attempt was made to obtain a blank of the available historic sample. A fragment from page 46 in "Kreuterbuch" without any visible application of copper pigment was scanned. Except for considerable amounts of copper ions, only minor amounts of calcium, magnesium and manganese in a magnitude of 10 times less than copper ions are detectable (figure 45). This observation shows very clearly the copper ions possess a remarkable mobility within the paper web and that they probably spread in a mm range away from their original application.

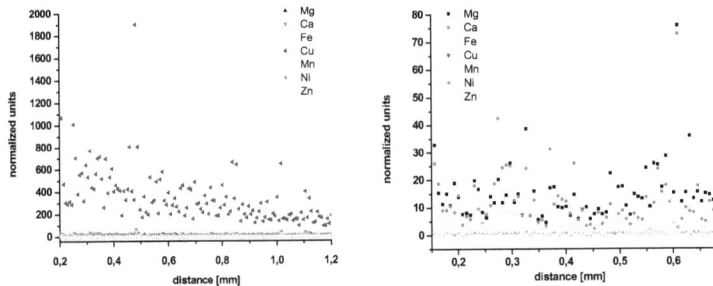

Figure 45. LA-ICP-MS analysis on page 46 of "Kreuterbuch". Left: A piece of sample was scanned that did not contain any visible application of copper pigment. Except for considerable amounts of copper ions no other ions are detectable in significant amounts. Right: Rescaled y-axis reveals some amounts of magnesium, manganese and calcium ions.

As the available sample was only about 3 mm long, it was scanned throughout the whole length of the fragment from blank to green pigment area. The amount of copper ions is increasing constantly towards pigment appliction. The rescaled y-axis reveals that the amounts of all other ions found in this sample are negligible compared to the amount of copper ions (figure 46). Contrary to irongall ink that only releases iron ions to a minor extent into the surrounding paper, copper ions are found throughout the paper as already indicated when scanning the blank sample. This observation may serve as an explanation why next to the pigment application and even in paper areas with no direct pigment contact considerable degradation of the cellulose is detected.

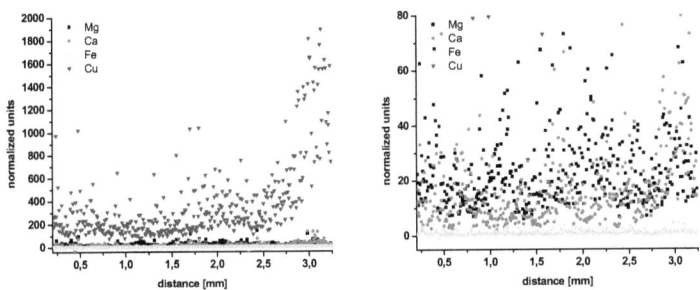

Figure 46. LA-ICP-MS analysis on page 46 of "Kreuterbuch". Left: Scan throughout the whole length of the fragment from blank to green leaf area. The amount of copper ions is increasing constantly. Right: Rescaled y-axis reveals some amounts of magnesium and calcium ions, slowly increasing towards ink application, but the amounts detected are in a magnitude of 10 times less than copper ions.

4.2.3 Treatment options

A main focus in the treatments section was placed on the investigation of the efficiency of aqueous calcium phytate/ calcium hydrogen carbonate treatment to inhibit irongall ink corrosion on model and historic papers. Several other treatments, like gelatine application to slow down irongall ink corrosion, and non-aqueous magnesium compounds with and without addition of antioxidants to inhibit degradation caused by copper ions were investigated, too.

Whatman filter paper (irongall ink corrosion, Kolbe)

Whatman filter paper was chosen to study the effect of ink corrosion and treatment on pure cellulose. As expected, non-treated paper with irongall ink application strongly degraded in the course of accelerated aging (figure 47). After 6 days of accelerated aging the average molecular weight of untreated inked paper decreased from 200 kg/mol to less than 100 kg/mol. This value is maintained even after six additional aging days. No further chain scission occurred; the decrease in average molecular weight has reached a constant level. Contrary to that, the total amount of carbonyl groups is increased throughout the whole aging period observed. In the beginning the untreated inked paper without aging has a carbonyl group content of 20 µmol/g. After 12 days of accelerated aging it increased to almost 70 µmol/g. In the case of total carbonyl groups no levelling off phase like for M_w was detected. As no further chain scission occurs between 6 and 12 days of accelerated aging and the amount of theoretical reducing end groups remains consequently stable at a value between 15 and 20 µmol/g, additional carbonyl groups can only be introduced by further oxidation as keto functionalities at C2 and C3 or as aldehyde functionality at C6. Aldehyde functionalities may be further oxidized to carboxylic acid. However, the development of carboxyl groups was not further analyzed. After treatment with calcium phytate/ calcium hydrogen carbonate a completely different development is observed. Only after 12 days of accelerated aging the average molecular weight is decreased. The development of carbonyl groups is equally inhibited. The difference between the total amount of carbonyl groups and the theoretical amount of carbonyl groups remains constant throughout the whole aging period. This indicates that besides the formation of additional reducing end groups originating from chain scission, no oxidation occurs.

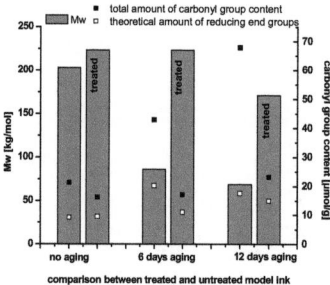

Figure 47. M_w, carbonyl group content and amount of reducing end groups in treated and untreated model ink covered areas without aging, after 6 days of aging and after 12 days of aging.

When plotting the increase of carbonyl group content versus aging time, the rate of formation of oxidized functionalities is visualized. For Whatman filter paper samples containing irongall ink that was treated with calcium phytate the slope is parallel to the slope of blank paper. This indicates a successful inhibition of oxidation. To underline the efficiency of the chosen treatment, samples with two ink modifications (OU-samples, unbalanced irongall ink, and OK-samples, unbalanced irongall inks containing additional copper ions) were compared to the paper blank and the treated sample. Untreated inks have a much steeper slope indicating faster oxidation (figure 48 left). Even though the addition of copper ions should accelerate the degradation, there is no difference between the two ink modifications. However, when subtracting the amount of reducing end groups, the effective amount of oxidized functionalities introduced by further oxidation can be evaluated. In figure 48 right, a difference between unbalanced ink (OU-samples) and unbalanced ink containing additional copper ions (OK-samples) becomes visible. The addition of copper ions leads to a more pronounced oxidation of the cellulose samples as can be seen in the steeper slope of OK-samples.

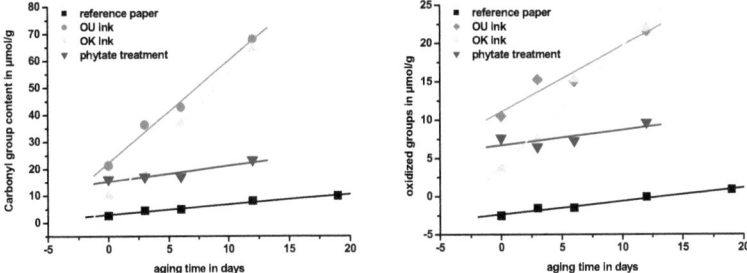

Figure 48. Development of carbonyl group content during accelerated aging. Left: Total amount of carbonyl group content. Right: Amount of oxidized groups without reducing end groups.

In figure 49 the M_w development of all different samples in the course of accelerated aging is compared. For irongall ink that was treated with calcium phytate the slope is even less steep than the slope of blank paper, indicating a clear beneficial effect of the treatment on the cellulose, too.

When gelatine is applied on Whatman filter paper without any further treatment it shows that gelatine has no pronounced stabilizing effect on cellulose itself when irongall ink (OU-sample) is present. The rates for oxidation as well as hydrolysis are in good agreement with the ink sample. Nevertheless after six days a tendency towards a levelling-off in molecular weight decrease can be observed. Even though gelatine sizing is far less effective than calcium phytate treatment, it leads to a low degree of protection towards chain scission. This effect shows a tendency to become more pronounced after prolonged aging.

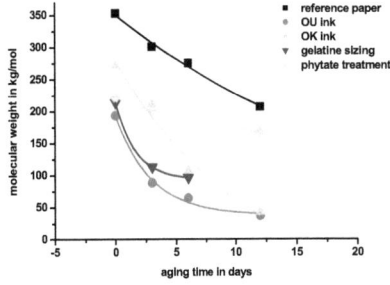

Figure 49. Development of molecular weight (M_w) during accelerated aging.

When comparing the initial molecular weight of the different samples, one important observation is that even before accelerated aging different values are obtained. The highest

M_w value is measured for the reference paper, followed by OK-samples (unbalanced ink containing copper ions) and OU-samples (unbalanced ink), with and without treatment. This observation, including the sequence of $M_{w\ reference}$ paper > $M_{w\ OK}$ > $M_{w\ OU}$, was already made in when studying the synergistic effect of single ink components, but on a different paper (model rag paper). Most chain scission obviously occurs when unbalanced ink is applied. The degradation is very fast, because also treated OU-samples have a similar M_w. This observation underlines that the relatively long period preparation and sample measurement is not the main influencing factor. The most probable explanation is a spontaneous degradation caused by ink application as already discussed in section 4.2.1.

Model rag (irongall ink corrosion)

A comparable set of treatment and aging was also performed on model rag papers with unbalanced, copper containing ink on it. Without treatment the expected shift in molecular weight distribution is observed (figure 50 left). As the MWD shifts uniformly, random degradation effecting equally low and high molecular weight regions is detected. There is no formation of a low molecular weight shoulder like observed in naturally induced irongall corrosion. The main degradation pathway is clearly oxidative indicated by the upward shift of $DS_{C=O}$-plots towards the DS_{REG}-plot. Unlike Whatman filter paper, instead of a levelling-off phase after one week of accelerated aging, there is rather an additional acceleration of degradation. After the calcium phytate treatment the MWD remains mainly unchanged, only after two weeks of accelerated aging a slight shift toward lower molecular weight can be observed. Oxidation is suppressed successfully, even after two weeks of accelerated aging hardly any difference in the $DS_{C=O}$-plots theoretical reducing end groups and the inked model rag paper can be observed (figure 50 right).

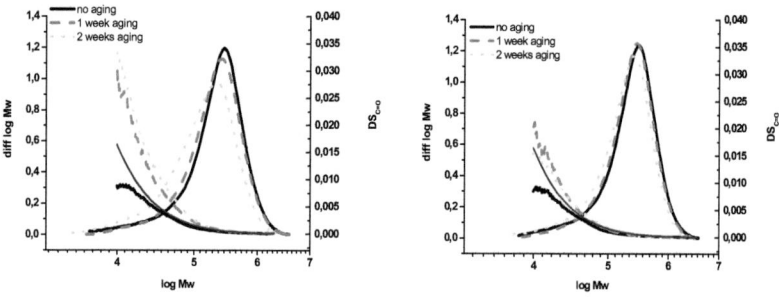

Figure 50. Left: Molecular weight distribution and degree of substitution ($DS_{C=O}$) plots for carbonyl groups of untreated model rag paper with unbalanced ink on it in the course of accelerated aging. Right: Molecular weight distribution and degree of substitution ($DS_{C=O}$) plots for carbonyl groups of treated model rag paper with unbalanced ink on it in the course of accelerated aging.

Like observed in the MWD of paper without ink application, there is no serious degradation of M_w and carbonyl group content. After one week no observable change in the parameters investigated took place and even after two weeks only minor changes occurred. When the paper was treated, there is trend towards higher stability of M_w (figure 51 left). The accelerated aging of the model rag paper with unbalanced ink applied on is more pronounced (figure 51 right). Average molecular weight remains quite stable after one week of aging. There is a small difference between treated and untreated ink areas right from the beginning, but only after two weeks of accelerated aging this difference becomes pronounced. The oxidation of treated paper covered with ink is successfully suppressed, no oxidation can be observed and the increase in carbonyl group content is due to some chain scission that occurred in the course of aging. In contrast to that, on untreated paper the difference between total amount and theoretical reducing end groups is increasing throughout the aging, indicating a strong oxidative influence as expected in irongall ink corrosion.

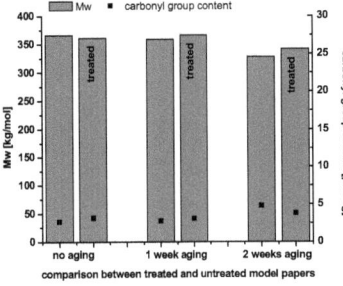 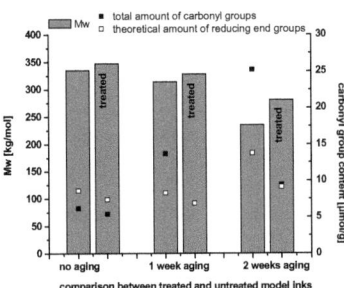

Figure 51. Left: M_w and carbonyl group content in treated and untreated model paper without aging and after two aging stages. Right: M_w, carbonyl group content and amount of reducing end groups in treated and untreated model ink covered areas without aging, after 6 days of aging and after 12 days of aging.

To conclude the results for model rag papers with model inks, there is a pronounced protective effect of calcium phytate treatment on the inked samples and a tendency towards protection of the cellulose in blank paper as well. The model paper cannot fully reflect naturally aged samples with irongall degradation; nevertheless, as unambiguously an oxidative damage was induced on the model papers, the protective effect of phytate treatments upon this type of degradation has been proven.

The chosen model rag papers are an important step towards the analysis of "real" paper, as these papers also contain hemicelluloses that considerably change the aging behaviour of paper. Compared to Whatman filter paper, the rate of degradation of cellulose obtained from hemp fibres is significantly slower; no levelling-off was achieved after two weeks of accelerated aging.

Historic papers with irongall ink ("K" samples)

Finally, historic rag papers with irongall ink were used to test the efficiency of calcium phytate/ calcium hydrogen carbonate on genuine material. Before accelerated aging the samples were analyzed to make sure that no unwanted side effects occur during phytate treatment. In figure 52 molecular weight distributions before and after treatment are shown. Neither plain paper nor inked papers suffer from any detectable change in molecular weight distribution after aqueous phytate treatment.

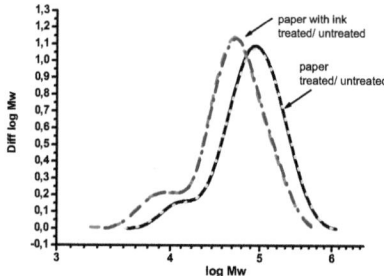

Figure 52. Historic "K" samples before accelerated aging. The aqueous phytate treatment has no immediate damaging effect on paper and on paper with ink.

The data obtained for the historic samples before and after treatment are summarized in figure 53. Even though the historic samples have been more oxidized at the beginning of the treatment than model papers, the inhibition of oxidation and hydrolysis can also be confirmed. On both, paper and inked paper, oxidation is suppressed and the molecular weight is stabilized nearly at the original level. Between the two aging steps no further degradation is observed. Both, carbonyl group content and M_w, have been stabilized close to their initial values.

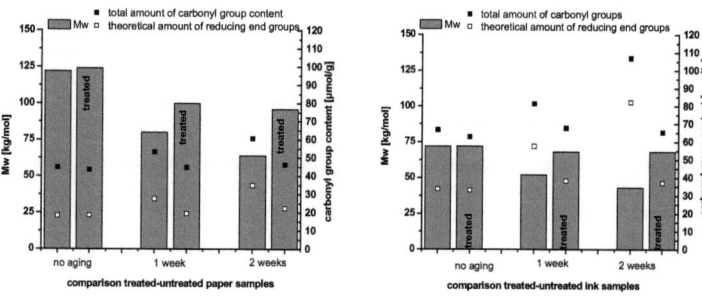

Figure 53. Left: M_w, carbonyl group content and amount of reducing end groups in treated and untreated paper after two aging stages. Right: M_w, carbonyl group content and amount of reducing end groups in treated and untreated ink covered areas without aging, after 6 days of aging and after 12 days of aging.

By means of LA-ICP-MS the distribution of selected ions on the paper surface can be followed. In figure 54 it can be observed that the historic ink of K-samples contains iron and copper ions, while calcium ions are only present in minor amounts. This is in agreement with previous studies [160]. After treatment the amount of detectable calcium ions has increased considerably. Their distribution follows copper and iron ion distribution. It is not yet known if the coexistence of Ca/Fe/P is caused by a chemical bond via bicomplexes of calcium and iron phytate. However the phytate represented by calcium ions obviously accumulates in close vicinity to the ink application, detected via iron and copper ion distribution. As the oxidative and hydrolytic degradation was proven to be inhibited, iron and copper ions were successfully hindered from entering the reaction system. Some of the copper and iron ions are probably washed out during aqueous treatment. However, as still considerable amounts were detectable after the treatment, the remaining iron and copper ions must have entered in a stable interaction with calcium phytate complexes.

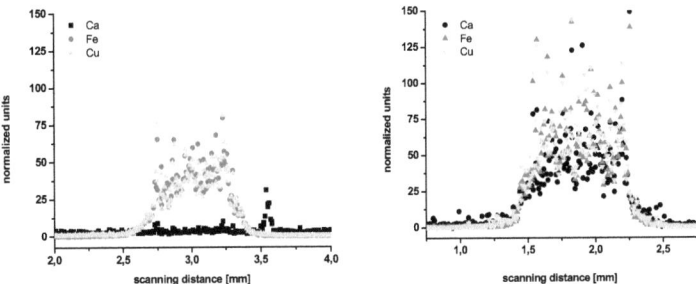

Figure 54. LA-ICP-MS scan across an inked areas of two different K-sample. The peak heights are different because the scans were performed on two different ink lines. Left: Before treatment. Right: After treatment.

Model rag papers + copper acetate

There are several indications that degradation of cellulose caused by copper ions is different from that caused by iron ions [99]. Consequently, for degradation caused by copper ions other treatment options have been sought for. Calcium phytate/ calcium hydrogen carbonate is very effective for the treatment of irongall ink corrosion, but its use in the context of degradation exclusively caused by copper ions is discussed controversially. There are publications about the adverse effects of phytate applications [147] and the opposite [161]. The studies leading to the conclusion that phytate treatment will encourage degradation caused by copper ions have not been published, while it was found that that the application of phytate was not detrimental to copper containing inks [108]. The example in the previous section of a historic rag paper covered with ink containing both, iron and copper ions, and the successful inhibition of ongoing degradation via calcium phytate treatment rather indicates that more research is needed in this topic to find out about the opportunities of calcium phytate treatment on corrosion caused by copper ions.

Application of magnesium compounds

Another option to slow down the activity of transition ions in general discussed in literature is the application of magnesium compounds. Their preventive actions towards oxidation caused by transition metal ions such as copper has been discovered in the context of bleaching in pulp production [122]. Also in investigations about copper containing pigments, it was found that when significant amounts of magnesium ions are present, the detrimental action of copper pigments was hindered [162]. Using the CCOA and FDAM-method to follow the development of oxidized functionalities and molecular weight this treatment approach was evaluated.

In table 21 the results for two treatments with solutions containing magnesium ions and no treatment are compared to sample material before aging. None of the tested options was able to stop the decrease in molecular weight and the increase in oxidized functionalities. A comparison of Δ C=O shows that both treatments containing magnesium compounds suppressed oxidative influences as compared to untreated sample material. This confirms the notion that magnesium salts inhibit oxidative degradation of cellulosics.

Table 21. Overview of data comparing aqueous and non-aqueous magnesium treatment with no treatment on copper pigment containing sample material after seven days of accelerated aging.

Treatment	M_w [kg/mol]	M_n [kg/mol]	C=O [µmol/g]	REG[1] [µmol/g]	Δ C=O[2]	COOH [µmol/g]	Surface pH
Aqueous Mg(HCO$_3$)$_2$	81	38	29.7	26.6	3.1	28.0	10.0
Non-aqueous magnesium alcoholate	124	52	19.5	19.1	0.4	18.9	9.6
Without treatment	110	58	26.8	17.4	9.3	18.4	7.3
Reference (no aging)	221	87	8.1	11.5	-3.4	17.4	7.1

[1] REG: Reducing end groups calculated from M_n (please note: this value assumes that reducing ends are not further oxidized and that determination of M_n is accurate)
[2] Δ C=O: Difference between measured carbonyl groups and calculated REG, index for oxidative influences

Still, the decrease of molecular weight impedes the conclusion that any of the two treatments could be recommended for the inhibition of degradation caused by copper containing pigments. Especially when the magnesium compounds were introduced by aqueous means chain scission plays an important role. The strong alkalinity realized after the treatment reveals the main drawback of applying aqueous just as well as non-aqueous magnesium compounds on the paper. Untreated paper had neutral surface pH that did not change significantly throughout the aging period. The two treatments based on magnesium compounds realized a surface pH between 10 and 10.5 on the model rag paper, which persisted throughout the aging period. The advantages of protection against oxidation by magnesium compounds are only slightly outweighed the drawbacks caused by high pH.

The loss in M_W can also be observed in the MWD. Aqueous magnesium hydrogen carbonate treatment caused a strong shift of molecular weight distribution to lower molecular weight fractions. Compared to the results without treatment, the degradation is rather enhanced than inhibited. When magnesium ions were introduced in a non-aqueous way the degradation is less pronounced than without treatment (figure 55).

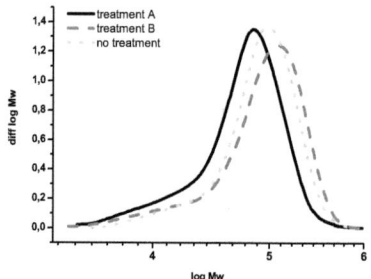

Figure 55. Molecular weight distribution of paper with copper pigment, submitted to different treatments with magnesium compounds after 7 days of accelerated aging at 80°C and 65% RH. Comparison of aqueous (A), non aqueous (B) and no treatment.

The determination of carboxyl groups gave further insight into possible reaction mechanisms. Carboxyl groups are mostly formed after β-elimination and by oxidation of aldehyde groups, but not during purely hydrolytic processes. Samples without treatment and after treatment B (non-aqueous) did not suffer further formation of carboxyl groups. After treatment A (aqueous) these groups increased significantly, underlining the importance of β-elimination after this treatment.

This is a very interesting observation because both substances realize an equally high pH, thus providing the condition for β-elimination. Usually, the application of water is considered to be beneficial for the conservation of paper as degradation products of cellulose and acids may be washed out. Especially for the partly soluble copper containing pigments a removal of noxious copper ions by aqueous washing has always been one aim of conservation treatments. In this context the aqueous application of magnesium compounds lead to an unexpectedly stronger degradation. One explanation for this negative result might be that instead of removing copper ions they were redistributed in the paper web, though this does not explain an increased damage on the analyzed samples that contained copper pigment anyway. Nevertheless, even a minor removal of copper ions should rather result in less than in more degradation. It may be concluded that aqueous removal of copper ions does not necessarily slow down the rate of degradation. Another explanation might be that due to a change of the water content in the paper the kinetics of degradation are influenced and changed. The main advantage of non-aqueous treatments is therefore the avoidance of water application in a reaction system strongly influenced by the presence of water.

Whether such non-aqueous solutions containing magnesium compounds should be employed on historic papers or not is determined by the oxidative pre-damage, due to alkaline degradation of oxycellulose under alkaline conditions. With copper ions present, which are active in alkaline conditions, cellulose is even more sensitive to degradation. Thus, the use of alkaline substances on oxidized cellulose must be seen rather critical. Furthermore, an alteration of pigments may occur, since some pigments are sensitive to pH changes. On the other hand, application of alkaline compounds on almost neutral paper without sizing and fillers will strongly increase the pH, while it will raise less on an acidic paper or on paper filled with pigment and size.

Application of antioxidants

As copper ions mainly cause oxidative damage on cellulose, the application of an antioxidant is also a promising approach. While pulp industry frequently works with antioxidants, paper conservators have generally been reluctant to apply them, mainly due to unwanted side effects like yellowing [163, 164].

When the chosen antioxidant, a mixture between methyl-*p*-hydroxybenzoate and propyl-*p*-hydroxybenzoate is applied in a low concentration, it revealed good results in inhibiting oxidation reactions after accelerated aging expressed in a superior preservation of MWD. Compared to that, no treatment and high concentration of the same mixture of antioxidants leads to stronger shifts in MWD (figure 56).

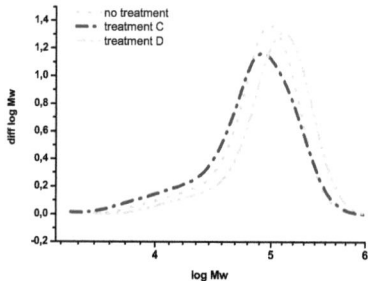

Figure 56. Molecular weight distribution of paper with copper pigment, submitted to different treatments with magnesium compounds after 7 days of accelerated aging at 80°C and 65% RH. Comparison of high (C) and low (D) concentration of antioxidant with no treatment.

The non aqueous treatment with low antioxidant content significantly reduced newly formed reducing end groups (formed hydrolytically or after β-elimination) and therefore resulted in a good overall performance. The opposite is true for non aqueous treatment with highly concentrated antioxidant. A decrease of molecular weight occurred even after application of non-aqueous magnesium alcoholates combined with antioxidant, but more moderate than after any other treatment (table 22).

Table 22. Overview of data comparing low and high concentration of antioxidants in non-aqueous magnesium treatment with no treatment on copper pigment containing sample material after seven days of accelerated aging.

Treatment	M_w [kg/mol]	M_n [kg/mol]	C=O [µmol/g]	REG[1] [µmol/g]	Δ C=O[2]
Low concentration	150	78	16.3	12.9	3.4
High concentration	104	44	28.8	23.0	5.8
Without treatment	110	58	26.8	17.4	9.3
Reference (no aging)	221	87	8.1	11.5	-3.4

[1] REG: Reducing end groups calculated from M_n (please note: this value assumes that reducing ends are not further oxidized and that determination of M_n is accurate)
[2] Δ C=O: Difference between measured carbonyl groups and calculated REG, index for oxidative influences

It is remarkable though, that a high concentration of antioxidants will cause rather negative results than improvements. One explanation might be the existence of a critical concentration. Passing this concentration might result in pro-oxidative effects, which are known for many (phenolic) antioxidants [165]. On the other hand the undesired reactivity of phenols with environmental pollutants and metallic impurities in polymers is more pronounced when larger amounts of antioxidants are present [166]. They might trap magnesium ions, thus impeding their protective efficiency.

4.2.4 Summary

An overview of results concerning the influence of transition metal ions on cellulose and paper is given in figure 57 and discussed in the following section.

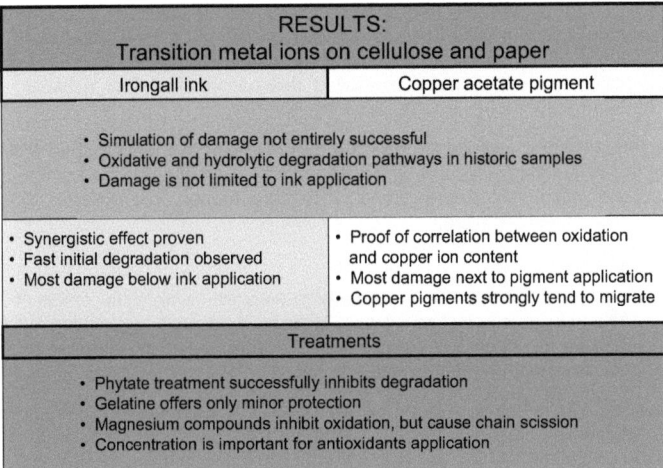

Figure 57. Overview of results from chapter 4.2 "Influence of transition metal ions on cellulose and paper".

Simulation of damage

Visual analysis and simple testing reveals that historic irongall ink lines exhibit a brownish colour and a tendency to absorb water in a preferred way. Ink lines are mechanically more fragile than the rest of the paper. Accelerated aging leads to a colour change on model inks. Most colour change and migration capacity was observed for tannic acid. The typical sorption behaviour of water on historic samples was not simulated in the chosen sample materials, neither in Whatman filter paper nor in model rag papers. The models remained hydrophobic. No preferred water sorption next to the ink line was observed. In Whatman filter paper, though, increased mechanical weakness was successfully inflicted during accelerated aging.

In chapter 4.1 a distinctive shoulder in the low molecular weight region of naturally aged papers was found to be one common feature. As it shows in almost all historic samples investigated in the context of degradation caused by iron and copper ions, it is considered to be typical of natural samples. Sub samples from areas with irongall ink or copper acetate applied directly onto it lead to an increase of this region. Thus degradation products accumulate. In some of the model papers the increase of this low molecular shoulder that is found in historic papers can be simulated during accelerated aging, especially when oxidative damage occurs due to irongall ink or copper pigments. Especially in Whatman filter paper this region gains significant importance. In this case all products found must originate from cellulose, being thus low molecular weight cellulose degradation products.

Like in simulated irongall ink samples, model rag papers with simulated degradation caused by copper ions showed a change in pigment colour after accelerated aging. In contrast to the irongall ink samples, it was possible to modify penetration properties of water. Further analysis revealed some degradation and formation of the low molecular weight shoulder that was found typical for aged and damaged papers. The main degradation pathway in these simulated samples was found to be oxidative occurring in the bulk material of cellulose. The model rag used for the simulation of degradation cause by copper pigment was different from the one used for irongall ink corrosion.

When analyzing the initial phase of irongall ink corrosion it was found that upon cycling humidity at relatively low temperatures no significant oxidative damage was achieved on all papers. Only when raising the temperature to 80°C the rate of oxidation increased considerably and within two days more oxidative degradation occurred than within one week of previous aging. Chain scission already started at the lower temperature in the initial phase. Therefore a successful simulation of oxidative degradation is rather achieved by higher temperatures around 80°C.

Another observation that was made during the analysis of the initial phase of irongall ink corrosion is a very fast degradation and oxidation that is probably induced by the application of the model ink without any further aging. This phenomenon was observed on Whatman filter paper and on model rag paper.

Degradation pathways

The study of ink modifications unambiguously proved the synergistic effect of single ink components. All parameters investigated were stronger influenced by unbalanced irongall ink than by balanced ink or single ink components, *i.e.* tannic acid or iron(II)sulphate. Tannic acid is instead acts rather as an antioxidant.

The analysis of historic sample material revealed that both, irongall ink corrosion and degradation caused by copper containing pigments may be driven by both, acid-catalyzed hydrolysis and oxidation. Oxidation is typically the prevailing pathway in degradation cause by transition metal ions, but hydrolytic degradation cannot be completely excluded.

All analyzed samples with degradation caused by copper ions showed the tendency to form trimodal molecular weight distributions, an observation that is generally not found in cellulose analysis of aged materials.

Spatial distribution of damage

Irongall ink corrosion and therefore degradation of cellulose is not always strictly limited to the area where the ink was applied on paper, but may extend to mm scale on adjacent paper. Nevertheless most damaged areas on all samples analyzed are always directly below the ink application.

Copper pigment containing samples do not necessarily have to be more degraded than pure paper samples. This observation is contrary to irongall ink corrosion where all ink covered parts have always suffered from more degradation than paper samples, even adjacent ones. On model papers it was demonstrated by LA-ICP-MS that copper ions diffuse into neighbouring paper areas upon accelerated aging and that the degradation in those areas was even more pronounced as directly underneath the pigment. Thus copper ions content was directly linked to oxidative damage of cellulose. On a historic sample material LA-ICP-MS proved that even on a blank of the available historic sample with no visible application of copper pigment, considerable amounts of copper ions are detectable.

Phytate treatment

The effect of phytate treatment on irongall ink covered pure cellulose was studied on Whatman filter paper. Stabilization of the molecular weight and carbonyl group development of the unbalanced irongall ink treated with the phytate solution was achieved successfully. The same is true for model rag papers. Oxidation of treated paper covered with ink is successfully suppressed, no oxidation can be observed and the increase in carbonyl group content is due to some chain scission that occurred in the course of aging. On historic samples oxidation is suppressed and the molecular weight is stabilized nearly at the original level. Additional analysis showed that neither plain paper nor inked papers suffer from any detectable change in molecular weight distribution after aqueous phytate treatment. After treatment the amount of detectable calcium ions has increased considerably as proven by LA-ICP-MS. Their distribution follows copper and iron ion distribution.

Gelatine treatment

Gelatine application on Whatman filter paper without any further treatment has only a low stabilizing effect on cellulose itself when irongall ink is present. However, recent investigations on the role of gelatine sizing came to the conclusion that gelatine acts as a buffer towards acids. Additionally, more moisture is retained by gelatine sized samples than by unsized samples [167]. This buffering capacity of gelatine would add to the buffering capacity generally found in historic rag papers originating from other additives, *e.g.* alkaline compounds used in the process of paper making. Opposite to that, more moisture content facilitates degradation processes in the paper web. In their study Baty and Barrett also suggested that the increased moisture gain observed under accelerated aging will decline upon natural aging. In this context, Whatman filter paper, an extremely pure material without fillers, submitted to relatively short accelerated aging is not an appropriate means to study the influence of gelatine on the stabilisation of degradation caused by irongall ink. Beneficial aspects of gelatine application contributing to more mechanical stability of the paper sheet are not directly reflected by GPC measurements because the gelatine sizing is removed prior to analysis.

Treatment with magnesium compounds

All deacidification treatments containing magnesium compounds suppressed oxidative influences as compared to untreated sample material. Unfortunately, especially aqueous magnesium hydrogen carbonate treatment decreased molecular weight considerably. When magnesium ions were introduced in a non-aqueous way samples showed better resistance against aging than untreated ones. Nevertheless, the advantages of protection against oxidation by magnesium compounds only slightly overrule the drawbacks caused by strongly alkaline pH.

Treatment with antioxidants

For antioxidant application the concentration was found to be crucial. When applied in low concentrations good results were achieved in inhibiting oxidation reactions in the course of accelerated aging. The opposite is true for highly concentrated antioxidant. One explanation might be that when passing threshold concentration pro-oxidative effects may gain importance.

4.3 Wet-dry interfaces

The formation of wet-dry interfaces (for preparation and sampling see figure 15 on page 53), often called tidelines, was studied on Whatman filter paper, unaged and aged (for aging procedure see page 51), and on historical papers with old water marks on it and freshly produced ones. Next to visual analysis sample material was investigated by CCOA and FDAM method for oxidized functionalities and molecular weight in cellulose. All results obtained after CCOA and FDAM labelling refer to changes exclusively on the cellulose molecule itself. Both labelling procedures do not detect water soluble degradation products or very low molecular weight compounds. Additionally analysis by LA-ICP-MS was included to study the distribution of metal ions within the paper and especially in the tideline region.

4.3.1 Wet-dry interfaces on Whatman filter paper

Preliminary analysis showed that the intensity of the desired tideline formation on Whatman filter paper is a function of time. The longer the paper stripes were left in water, the more pronounced the obtained tideline was. In the context of the influence of transition metal ions on cellulose it is agreed upon that these ions will cause oxidative degradation on cellulose. This degradation is generally accompanied by discolouration of paper, usually browning in the close vicinity of the pigment or ink. In order to find out about the causes of tideline formation also the influence of metal ions was investigated. Most of the artificial tidelines were produced on Whatman filter paper. As outlined in the materials and methods section, this cellulosic material is expected to contain only very low amounts of impurities. Analysis by LA-ICP-MS confirmed that Whatman filter paper as purchased does not contain any significant amounts of transition metal ions of interest (figure 58). A 0.5 mm scan on the paper surface shows only minor amounts of calcium ions.

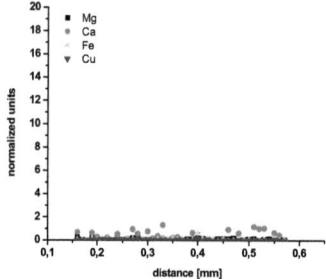

Figure 58. LA-ICP-MS scan on Whatman filter paper blank as purchased.

Additional investigations were performed using a confocal laser scanning microscope (CLSM) to detect fluorescence on the paper surface. A sharp border between fluorescent and non-fluorescent paper surface was found when a tideline produced on Whatman filter paper was analyzed by CLSM (figure 59). The origin of this fluorescence is unknown since no specific labels have been used.

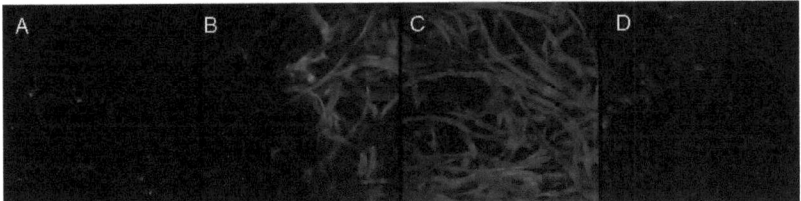

Figure 59. Fluorescent wet-dry interface. A: above tideline (not immersed into water during experiment), B: sharp fluorescent boarder, C: within the tideline region, D: below tideline (immersed in water during the experiment), some slight fluorescence is still visible.

When the Whatman filter paper was washed in water prior to further analysis the brown tideline was removed. The fluorescence was also removed as seen in figure 60. Obviously these fluorescent substances are water soluble, at least within short periods of time like several weeks. This leads to the conclusion that rather than oxidized cellulose water soluble products will cause fluorescence in the wet dry interface. Another suggestion is that the fluorescent species are only formed upon immersion in water.

Figure 60. Paper sample like in figure 59, after immersion in water, three pictures were taken on different positions on the paper, including former tideline region, no fluorescence can be detected.

Further analysis by CCOA and FDAM aimed towards quantification of oxidation and hydrolysis in cellulose if present. All measurements have been performed on samples that have been produced under lab conditions without any special precautions.

In table 23 all results for carbonyl group determination of Whatman filter paper are summarized. The measured amount of carbonyl groups in cellulose is generally very low in all reference samples, and there is no significant difference between the values. The same is true for molecular weight. Therefore, an immediate measurable influence on cellulose is excluded after tideline creation. After completing the accelerated aging period (5d), a decrease in molecular weight of the "below" sample is observed. The decrease in the "below" sub sample is more pronounced than in "above" and "tideline" sub samples. The later ones are in a comparable condition. Chain scission seems to start even a little bit later in "tideline" sub sample as after two aging days no loss in molecular weight is observed. Also, when comparing carbonyl group content, "tideline" and "above" are in a similar condition, containing two times more carbonyl groups. In the "below" sample, there are three times more carbonyl groups.

Table 23. Data overview of Whatman filter papers after CCOA-labelling (batches cs77 for reference and cs86 for aged samples).

Sample ID	M_w [kg/mol]	M_n [kg/mol]	M_z [kg/mol]	PDI[1] M_w / M_n	C=O [µmol/g]
reference below	295	129	54	2.27	2.6
reference tideline	294	121	555	2.42	2.4
reference above	292	115	551	2.54	2.1
2d aging below	267	139	470	1.93	3.8
2d aging tideline	301	182	467	1.67	2.6
2d aging above	271	126	476	2.14	2.9
5d aging below	175	77	321	2.28	8.7
5d aging tideline	207	93	365	2.21	4.5
5d aging above	223	94	450	2.38	4.2

[1] PDI: Polydispersity index: M_w / M_n

In table 24 all results for carboxyl group determination of Whatman filter paper are summarized. The measured amount of carboxyl groups is generally low in all reference samples, and in the reference material there are only small differences between these values. According to FDAM M_w determination, the "above" sample is in a slightly better condition than "tideline" and "below". This relation remains after completing the aging period (5d): "above" has a higher M_w than "below", worst results are obtained for "tideline".

Table 24. Data overview of Whatman filter papers after FDAM-labelling (batch COOH79).

Sample ID	M_w [kg/mol]	M_n [kg/mol]	M_z [kg/mol]	PDI[1] M_w / M_n	COOH [µmol/g]
reference below	355	183	605	1.94	5.5
reference tideline	343	166	545	2.07	6.0
reference above	393	248	592	1.58	5.4
2d aging below	271	151	434	1.80	5.1
2d aging tideline	279	162	456	1.73	6.6
2d aging above	332	189	539	1.75	5.3
5d aging below	220	107	377	2.06	6.7
5d aging tideline	176	81	297	2.16	5.1
5d aging above	276	171	426	1.62	5.3

[1] PDI: Polydispersity index: M_w / M_n

The label designed for the FDAM-method mainly detects uronic acids that develop via aldehyde structures in cellulose. As no pronounced amount of these structures was detected by CCOA-analysis, no increased carboxyl group content is expected. This is reflected by the analysis: the values obtained vary from 5.1 µmol/g to 6.7 µmol/g, only a very small range of 1.6 µmol/g is covered by the samples analyzed. For the "above" sample the values remain unchanged throughout the accelerated aging period. This is in agreement with previous analysis in chapter 4.2 when during accelerated aging of Whatman filter paper only acid hydrolysis was detected. The interpretation of "below" and "tideline" samples is more difficult. While in the first two steps (no aging and two days of aging) a tendency towards increased carboxyl group formation in the "tideline" sample can be detected, this is not true any longer after five days of aging. In the "below" sample there is a slight trend towards increase of carboxyl groups. This correlates with its increased amount of carbonyl group content. A graphic overview is given in figure 61 left.

The molecular weight obtained by FDAM-labelling of the samples is higher than after CCOA-labelling. This observation is quite unusual according to lab experience. In fact, the correlation between M_w determination after CCOA-labelling and M_w determination after FDAM-labelling is lower than expected with an R value of 0.8. Nevertheless, it should be kept in mind that two different sub samples were analyzed in two different runs. These two values can be considered as double determination. When plotting the average of them there is a significant trend towards least degradation in the "above" sample, while "below" and "tideline" are equally degraded (figure 61 right).

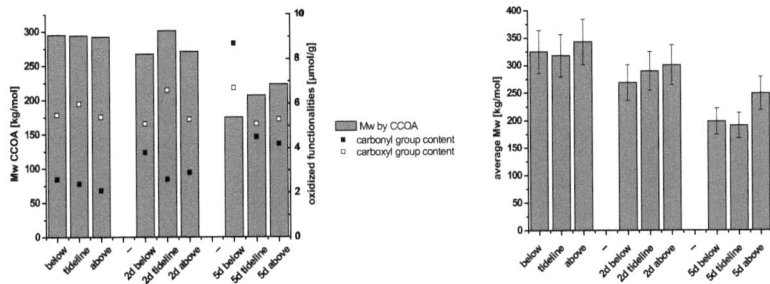

Figure 61. Overview of results of artificial tidelines on Whatman filter paper. Left: M_w, carbonyl and oxidized functionalities. Right: M_w of „below", „tideline" and „above" sample with calculated standard deviation (SD = 12%, n = 2).

Regarding the molecular weight distribution of unaged samples, it is uniform in all three samples "above", "below" and "tideline" after CCOA-labelling (figure 62 left). Thus, the formation of a visible brown line does not induce any immediate changes in the molecular weight distribution. The same is true for the fluorescence signal (figure 62 right).

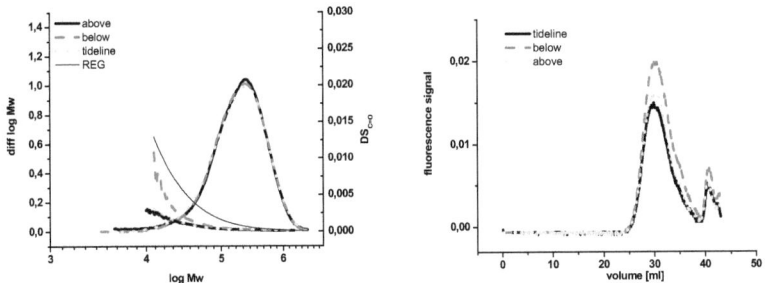

Figure 62. Whatman filter paper with artificial tidelines. Left: The differential molecular weight distributions and the $DS_{C=O}$-plots do not detect any difference between the three sub samples. Right: The fluorescence signal of all three sub samples exhibits the same shape.

After artificial aging for five days, only minor changes in the MWD can be detected. There is obviously no significant oxidation, because the plotted carbonyl groups are still well below the calculated amount of reducing end groups as can be seen in carbonyl group profiles in figure 63 left. The profile of the "below"-plot is slightly increased towards the two other samples, but still does not exceed the DS_{REG}-plot. An interesting observation is made when comparing the fluorescence signals of the three sub samples after five days of accelerated aging (figure 63 right). During accelerated aging a shoulder develops in the "tideline" and "below" sub samples that is not found in the "above" sample. This shoulder corresponds to low molecular weight substances, indicating the molecular weight range that is affected by changes due to water immersion and tideline formation. The difference that leads to the appearance of this shoulder cannot be quantified by carbonyl group determination, "above" and "tideline" have roughly the same value (4.2 µmol/g respectively 4.5 µmol/g).

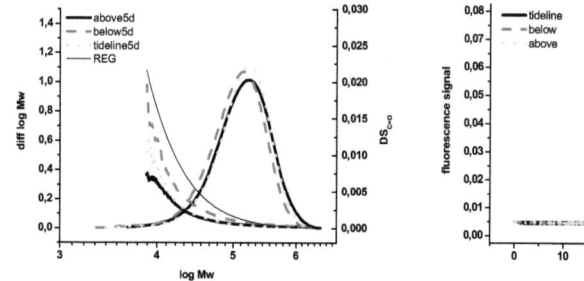

Figure 63. Whatman filter paper with artificial tidelines after five days of accelerated aging. Left: Differential molecular weight distribution after CCOA-labelling. Right: In the fluorescence signal shoulder develops in the "tideline" and "below" sub sample (see arrows) that is not detected in the "above sample". This change in fluorescence signal cannot be quantified in the total carbonyl group content.

Whatman filter paper does not contain any significant amounts of metal ions as shown on blank paper by LA-ICP-MS. After tideline formation through immersion into demineralised water a clear increase in metal ions especially in the tideline region can be detected (figure 64 left). Especially the increase in copper ion content is remarkable. As this test paper was produced under normal lab conditions it was assumed that copper ions have been introduced into the paper from the lab environment, either through contaminated water or other equipment used to produce the tidelines.

In order to avoid any influence from the environment another set of tideline samples on Whatman filter paper were produced under clean room conditions. The characteristic brown line will appear, too, and so does fluorescence, but as can be proven by LA-ICP-MS, only calcium and magnesium ions will be increased significantly in the tideline region (figure 64 right).

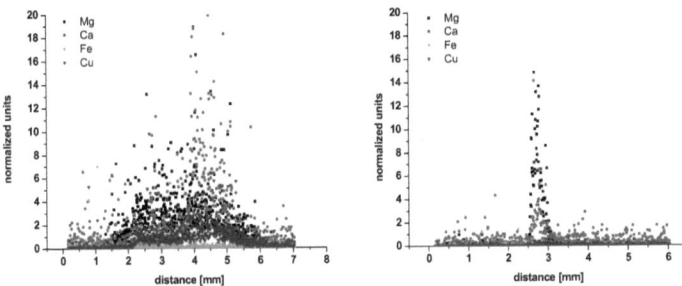

Figure 64. Both Whatman filter papers with tide line region scanned. Left: The tideline was produced under lab conditions with demineralised water. Right: The tideline was produced under clean room conditions with bi-distilled water.

On the samples analyzed an interesting increase in copper, magnesium and calcium ion content is observed in the tideline region. While the copper ions most probably originate from the lab environment, magnesium and calcium ions are contained in the Whatman paper itself and transported to the tideline region. Copper ions are powerful catalysts to accelerate oxidative degradation on cellulose as shown in chapter 4.2.2, but no pronounced degradation was shown in the "tideline". When comparing the copper ion content in the tideline region to the content found in an historic paper it is still extremely low. This might explain why copper ion induced oxidation does not play an important role in the context of tideline samples.

Still, an explanation for the formation of a brown line and fluorescence is missing. The content of magnesium ions might be an interesting hint in this context. It has been discussed many times, that deacidification based on magnesium compounds is not advised any longer, because it will lead to more discolouration of the treated paper than calcium based treatments (for further discussion see chapter 4.4 Mass deacidification). The mechanism behind the observed discolourations has not been solved yet; nevertheless, as in Whatman filter paper the accumulation of a brown line is always observed even when working under extremely clean conditions, it might be the same phenomenon.

4.3.2 Wet-dry interfaces on historic papers

When analyzing historic papers (see page 48) one major disadvantage is that the source of the noticeable tideline is not known. It can only be assumed that the tideline was produced under the influence of water. Even when agreeing upon water as the reason for a tideline it remains unclear whether clean or dirty water happened to enter the paper and in which way it was applied. Therefore, generalization about degradation patterns in historic papers with pronounced tidelines is rendered extremely difficult. This general remark is underlined when studying the results in table 25 that summarizes all results for carbonyl group determination in historic rag papers. All values obtained are in a range that may be expected for historic rag papers according to the findings presented in chapter 4.1. None of the analyzed samples supports the conclusion put forward that more degradation can be found in the tideline region, as M_w of "tideline" either equals or exceeds the value of M_w found in "below" or "above" sub samples. Only one sample (H_88) supports the thesis that more oxidation is found in the "tideline" region. Contrary to Whatman filter paper samples, most often lowest carbonyl group contents are found in "below" samples. Nevertheless, the differences are usually very small. When studying the results for M_w in the different sub samples of historic rag papers it shows again, that a general trend is hard to define. In most cases, "below" has lowest M_w which is in agreement with findings from model papers.

Table 25. Data overview of historic rag papers after CCOA-labelling (batch cs85).

Sample ID	M_w [kg/mol]	M_n [kg/mol]	M_z [kg/mol]	PDI[1] M_w/M_n	REG[2] theor.	C=O [µmol/g]
H88_below	188	69	422	2.73	14.5	14.7
H88_tideline	219	43	459	5.06	23.1	21.8
H88_above	213	71	427	3.02	14.2	16.8
H111_below	257	92	527	2.81	10.9	13.5
H111_tideline	246	71	474	3.49	14.2	11.2
H111_above	239	96	476	2.49	10.4	13.4
H112_below	203	73	409	2.79	13.7	14.3
H112_tideline	208	62	435	3.34	16.1	16.2
H112_above	235	83	487	2.84	12.1	17.3
H99_below	145	32	349	4.49	31.1	19.2
H99_tideline	171	42	411	4.09	23.9	24.4
H99_above	179	49	397	3.66	20.5	24.1
H95_below	253	81	602	3.15	12.4	16.6
H95_tideline	253	60	573	4.20	16.6	15.5
H95_above	222	69	480	3.22	14.5	14.4

[1] PDI: Polydispersity index: M_w/M_n
[2] REG: Reducing end groups calculated from M_n (please note: this value assumes that reducing ends are not further oxidized and that determination of M_n is accurate)

After FDAM-labelling, none of the analyzed samples supports the conclusion put forward that more degradation can be found in the tideline region, as the M_w of "tideline" either equals or exceeds the values for M_w of "below" or "above" (table 26).

Regarding carboxyl groups, there are more oxidized functionalities in the "tideline" region of historic rag, which is in agreement with other publications [134]. Nevertheless the differences in carboxyl group content are not very pronounced, except for sample H_88. In accordance with the findings after CCOA-labelling, most often lowest carboxyl group contents are found in "below" samples. This is again not true for model papers. Results for M_w in the different sub samples of historic rag papers show again, that a general trend is hard to define. In most cases, "below" has lowest M_w which is in agreement with findings from model papers.

Table 26. Data overview of historic rag papers after FDAM-labelling (batch COOH75).

Sample ID	M_w [kg/mol]	M_n [kg/mol]	M_z [kg/mol]	PDI[1] M_w / M_n	COOH [µmol/g]
H88_below	147	43	356	3.43	15.9
H88_tideline	219	64	461	3.43	21.0
H88_above	232	67	467	3.46	12.9
H111_below	287	115	588	2.48	11.8
H111_tideline	295	102	612	2.88	14.6
H111_above	269	90	564	3.00	11.3
H112_below	154	35	365	4.38	17.8
H112_tideline	202	70	422	2.89	19.9
H112_above	236	76	490	3.09	18.5
H99_below	160	53	408	3.00	16.0
H99_tideline	205	64	466	3.19	19.9
H99_above	203	61	477	3.32	18.6
H95_below	235	62	544	3.77	15.7
H95_tideline	301	81	695	3.70	15.9
H95_above	202	59	455	3.40	17.1

[1] PDI: Polydispersity index: M_w / M_n

As for tidelines on Whatman filter paper, M_w derived from CCOA and from FDAM-labelling may be considered as double determination (figure 65). In samples H_111 there is no significant difference between the different sub sampling areas at all, in most samples "below" has suffered from more degradation than other sub sampling areas.

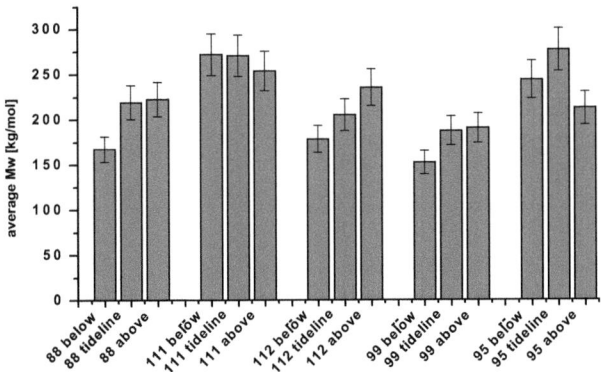

Figure 65. Overview of results of tidelines on historic rag papers. M_w of „below", „tideline" and „above" sample with calculated standard deviation (SD = 8.5 %, n = 2).

The extent of oxidation is analyzed by comparing the total amount of carbonyl groups with the theoretically calculated amount of reducing end groups (figure 66 above). In most cases theses numbers are quite equal. Again, no general trend is observed. Figure 66 below gives an overview of M_w determination after FDAM-labelling, including both, carbonyl and carboxyl groups.

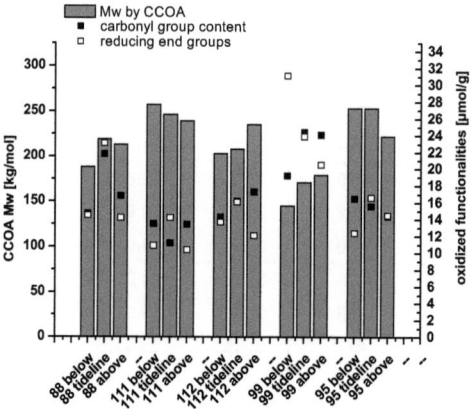

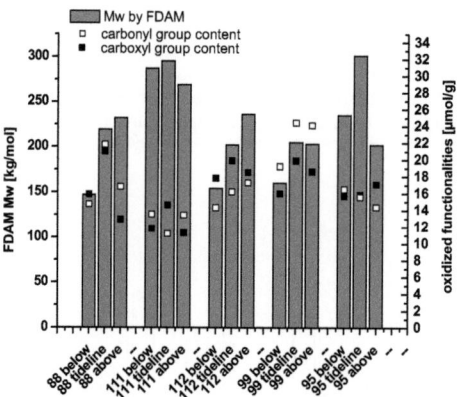

Figure 66. Overview of results of tidelines on historic rag papers. Above: M_w after CCOA-labelling of "below", "tideline", and "above" samples with total carbonyl group content and calculated reducing end groups. Below: M_w after FDAM-labelling of "below", "tideline", and "above" samples with carbonyl and carboxyl group content.

Molecular weight distributions reflect the findings presented above. Sample H_111 (no significant changes in M_w) and sample H_99 (significant differences between "below" on the one hand and "tideline" and "above" on the other hand) have been chosen to represent the phenomena observed on tidelines on historic rag samples. In sample H_99, there is a clear trend that "below" is most degraded (figure 67). The chain scission has affected also high molecular weight regions. Contrary to that in sample H_111 the distributions run almost identical in all molecular regions (figure 68). None of the selected sample MWDs support the thesis of significant oxidation to carbonyl groups as all $DS_{C=O}$-plots run well below the DS_{REG}-plot.

Interestingly, after FDAM-labelling the low molecular weight shoulder that was defined to be typical for historic papers is detected more clearly than after CCOA-labelling (figures 67 and 68, both right).

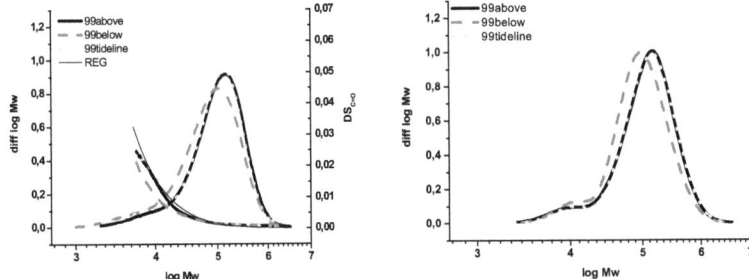

Figure 67. Differential molecular weight distribution of sample 99. Left: after CCOA-labelling. Right: after FDAM-labelling

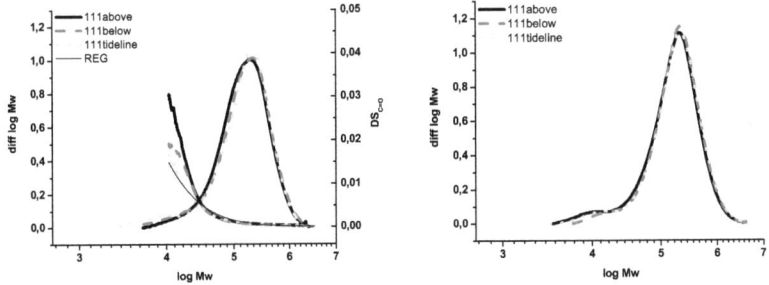

Figure 68. Differential molecular weight distribution of sample 111. Left: after CCOA-labelling. Right: after FDAM-labelling

Up to 100 times more ions are detected by LA-ICP-MS in the historic rag samples than in Whatman filter paper. Interestingly, the distribution of ions does not form a clear tideline peak like observed in Whatman filter paper (figure 69 left). Calcium ions have some tendency to accumulate in a very small area, but not in a very pronounced way. Contrary to that an accumulation of ions can be found in historic rag paper after producing a new tideline by application of water. Magnesium and calcium ions will be transported as well as manganese ions that are present in significant amounts in this specific paper (figure 69 right).

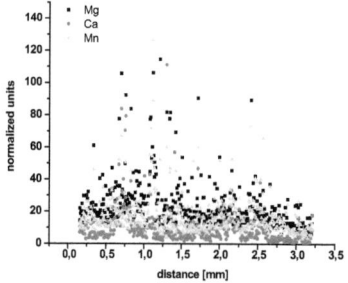 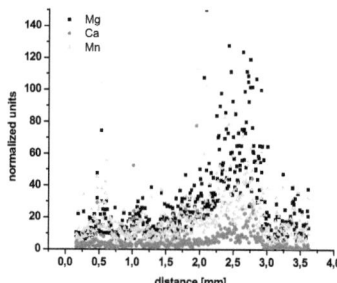

Figure 69. LA-ICP-MS analysis on rag paper H_112. Left: Scan across an historic water mark found on the paper of unknown origin. Right: Scan across a freshly produced tideline on historic rag paper using demineralised water.

In this specific sample there is an obvious difference between historic tidelines and freshly produced ones when it comes to the distribution of ions. One possible interpretation of this observation is that the historic mark found on paper H_112 does not originate from liquid water. Another interpretation is that during natural aging further de- and adsorption of humidity caused a re-distribution of ions within the paper web. As a short application of one droplet of water is enough to encourage significant migration this interpretation seems reasonable as it points to the high mobility within the paper web of these ions in general.

4.3.3 Summary

Neither Whatman filter papers nor historic papers evidence a locally restricted increase in chain scission of the cellulose when analyzing the so called "tideline" region. Instead, most degradation of the cellulose is generally observed in the "below" region, *i.e.* the part of model paper that was dipped into water or the part of the historic paper from where water accessed respectively. There are exceptions to this observation, especially in historic samples. Some of them did not experience any different degradation at all.

Another interesting observation is a difference in FDAM and CCOA procedure regarding the low molecular region. While samples analyzed according to the CCOA procedure only show increased low molecular weight region in the "tideline" samples, this is not true for the samples treated according to the FDAM procedure. In these samples there is always a pronounced low molecular weight region. This observation might be a hint for accumulating low molecular weight fractions in the "tideline" sample. In model papers this accumulation is not observed at all, neither in FDAM nor in CCOA procedure. Some contradicting results, *e.g.* the increase in molecular weight in sample "tideline" after two days of aging, might be due to inhomogeneities during manual sample preparation.

Often, the tideline phenomenon is related to oxidation, especially carboxyl groups are expected to increase after aging. According to the present experiments it can clearly be stated, that there is generally no increase in carbonyl groups, with the exception of sample H_88. In model papers the "below" area has more carbonyl groups, but this observation is not found in historic papers. Generally, the trend for carbonyl group development in historic papers is not very clear, sometimes "below" has the lowest carbonyl group content, sometimes the highest content can be found in this region. With carboxyl group content, there is no clear trend in model papers, the amount stays more or less unchanged along the aging period. Looking at historic papers, in most of the cases increased carboxyl group content is found in the "tideline" region, supporting observations made before, that methylene blue will stain the tideline region more intensely. Nevertheless, as generally only minor differences were observed applying the FDAM-labelling (within a range of 1.3 to 3.3 units) it is questionable if simple methylene blue staining is able to detect these small differences. Additionally, the methylene blue method for carboxyl group determination is very sensitive towards pH changes. Therefore, an alkaline buffer has to be used for quantitative analysis. Probably differences in colour intensity after methylene blue staining might just as well be caused by interferences like different surface pH or paper components.

An important finding is that in model papers the amount of metal ions introduced during the creation of the tideline is not very well controlled. To circumvent the introduction of all kinds of metal ions during tideline preparation work was performed under clean room conditions. The introduction of foreign metal ions was inhibited. Nevertheless, the accumulation of magnesium and calcium ions is a reproducible occurrence that is not influenced by external sources. In historic tidelines metal ions are not necessarily accumulated like in model papers. These observations lead to the general question in how far model papers with deliberately introduced tidelines will represent historic wet-dry interfaces. The question arises in how far model studies on Whatman filter paper rather observe changes made in metal ion content externally introduced than changes that are caused by cellulose degradation products.

A possible explanation for the formation of the brown line that defines the tideline region in Whatman filter paper is the accumulation of magnesium ions. Especially in the context of aqueous paper deacidification the use of magnesium compounds is not recommended as it is reported to lead to increased discolouration of the treated paper. The mechanism behind the observed discolorations has not been solved yet; nevertheless, as in Whatman filter paper the accumulation of a brown line is always observed even when working under extremely clean conditions, it might be the same phenomenon.

The phenomenon of fluorescence is still not clarified. From the present results it can be concluded that it is not caused by oxidation of cellulose. As GPC is not able to detect oligomeres and smaller degradation products from cellulose further analysis in this topic is needed.

4.4 Mass deacidification

The purpose of mass deacidification is the treatment of acidic paper on a large scale with the aim to neutralize present acids and at the same time to deposit an alkaline reserve to neutralize acids formed in the future. To achieve this aim, deacidification in general and especially mass deacidification systems use compounds that induce a strongly alkaline environment on degraded and pre-oxidized papers. This might cause β-elimination, a degradation reaction that leads to chain scission of the cellulose molecule. Fluorescence labelling followed by GPC can be used to study to probability of β-elimination. Sustainability of mass deacidification treatments has to be analyzed to prove the benefit of the treatment over possible disadvantages like β-elimination.

Next to an increase of pH above 7 and the deposition of an alkaline reserve, homogeneity, monitored by the deposition of magnesium ions, is another parameter that defines the quality of the deacidification treatment. In some cases it was observed that treated books have not been deacidified homogeneously, *i.e.* some regions within one sheet remain acid while others have reached an alkaline pH as intended. One possible explanation of this observation is the inherent inhomogeneity of historic paper materials as demonstrated in chapter 4.1. The effect of an inhomogeneous increase in surface pH on the development of molecular weight as a very sensitive indicator for degradation of the cellulose molecule is studied in the course of accelerated aging.

4.4.1 β-elimination

On historic papers, there has not only been hydrolytic degradation caused by acids but in the course of time also oxidized functionalities have been introduced. Deacidification treatments, especially non-aqueous ones introduce treatment liquids into the paper that have the potential to form strongly alkaline products. Prior to the immersion with the deacidification liquid the amount of water absorbed on the paper is usually reduced as most processes take place in an organic solvent system [136, 168]. Depending on the type of treatment chosen, the paper was either dried before treatment or cooled to slow down reactions. Nevertheless, successful deacidification treatments include a re-introduction of water, generally called reconditioning. Upon this re-adsorption of water the formation of the alkaline reserve via reaction with the deacidification agent takes place [136, 168]. During this desired reaction, strong alkalinity may adversely influence the stability of paper as alkaline degradation may occur, namely β-elimination. Under acidic conditions this reaction pathway is of inferior importance but the more alkaline the conditions the more important this reaction. It will lead to the scission of cellulose molecules resulting in rapid loss of mechanical stability.

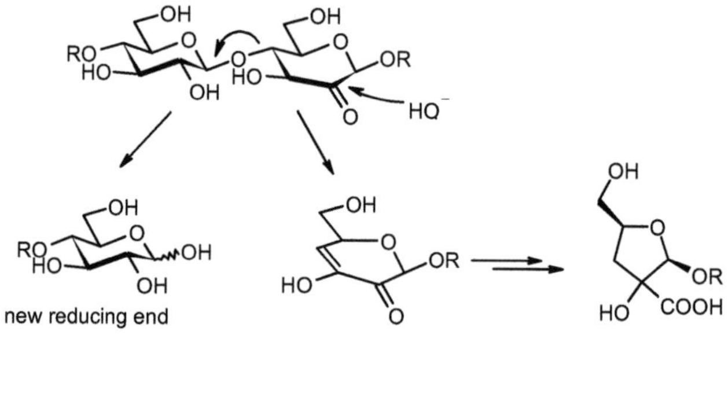

Figure 70. β-elimination of cellulose, R= cellulose chain

During β-elimination the amount of carbonyl groups is not decreased because for every cleaved chain, a new reducing end group is produced (figure 70). Contrary to that the amount of carboxyl group should be increased as rearrangement reactions yield a carboxyl group [169]. In order to study the likelihood of β-elimination three different pulps have been used as test material:

- native pulp (BKZO, bleached beech sulphite pulp, see page 45)
- the same pulp after sodium hypochlorite treatment at pH 7 for 30 minutes
- the same pulp after sodium hypochlorite treatment at pH 7 for 60 minutes

The sodium hypochlorite treatment was performed to introduce an increasing amount of carbonyl groups (see page 49). This is intended to increase the susceptibility towards β-elimination. The native pulp and the two portions of oxidized pulps were than processed as lab hand sheets and subjected to mass deacidification treatment according to the CSC booksaver© process.

Native pulp

The native pulp had a molecular weight of 314 kg/ mol (CCOA) respectively 300 kg/mol (FDAM), 17.5 µmol/g carbonyl group content and 18.0 µmol/g carboxyl group content. In the first rows of tables 26 to 28 the expected changes of the native pulp due to oxidation can be observed. Molecular weight decreased from its initial value down to less than 200 kg/mol after an oxidation treatment of 60 minutes, while oxidized functionalities increase as intended. Carbonyl group content increases to more than 40 µmol/g. Carboxyl group changes are not that pronounced: they increase from 18 µmol/g to 23 µmol/g. For M_w after FDAM labelling always lower values are observed, but both determinations correlate with an excellent value of R = 0.99.

Table 27. Overview of results for native pulp without sodium hypochlorite treatment

	M_w CCOA [kg/mol]	M_w FDAM [kg/mol]	C=O [µmol/g]	COOH [µmol/g]
No treatment	314	300	17.5	18.0
Run A	320	290	18.4	16.6
Run B	305	297	18.4	22.3
Run C	313	281	17.6	17.5
Run D	293	307	17.9	16.9
Run E	322	301	18.2	17.5
Average [runs A - E]	311	295	18.1	18.2
SD [unit]	12	6	0.4	2.4
SD [%]	4	2	2.0	13.0

Pulp hand sheets with an initially high molecular weight and low (< 20 µmol/g) carbonyl group content are not influenced by deacidification treatments. The amount of oxidized functionalities, molecular weight and molecular weight distribution do not change significantly (table 27). Thus no β-elimination was induced. For the treated pulp hand sheets the $DS_{C=O}$-plot is always slightly above the $DS_{C=O}$-plot of untreated samples (figure 71). This indicates some minor oxidation occurring in lower molecular weight region during deacidification treatment. Another small change that can be observed is a slight shift towards lower molecular weight regions. Some smaller fragments get lost during the treatment, the distribution gets slightly more narrow.

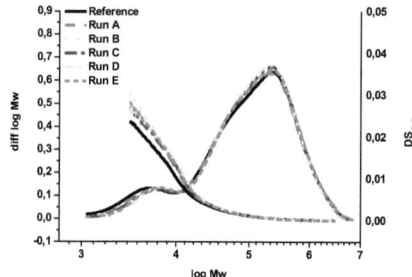

Figure 71. Molecular weight distribution and carbonyl group profile of the reference sheets without sodium hypochlorite treatment. No decrease of MW can be detected in the MWD, therefore no significant degradation processes do not take place.

Oxidized pulps

An oxidation treatment of 30 minutes in sodium hypochlorite at pH 7 leads to a remarkable decrease in M_w and to an increase of carbonyl groups by the factor of 2 (tables 27 and 28). Carboxyl group content only shows minor changes. After deacidification treatment of oxidized pulp hand sheets some degradation can be observed. M_w is up to 30 kg/mol lower than in non deacidified samples, exceeding clearly the standard deviation (table 28). The same is true for the increase in carbonyl group content. Even though some degradation is clearly observed, no significant increase in carboxyl group content has occurred so far that would clearly indicate β-elimination.

Table 28. Overview of results for pulp samples with a carbonyl group content of 34.5 µmol/g (30 minutes of sodium hypochlorite treatment)

	M_w CCOA [kg/mol]	M_w FDAM [kg/mol]	C=O [µmol/g]	REG[1] [µmol/g]	COOH [µmol/g]
No treatment	219	194	34.5	27.2	19.7
Run A	187	158	29.0	31.0	20.8
Run B	201	156	31.0	26.3	19.6
Run C	199	160	30.3	38.3	21.5
Run D	185	168	31.8	24.2	21.6
Run E	191	164	30.7	27.7	19.2
Average [runs A – E]	193	161	30.5	29.5	20.6
SD [unit]	7	5	1.1		1.1
SD [%]	4	3	3.5		5.4

[1] REG: Reducing end groups calculated from M_n (please note: this value assumes that reducing ends are not further oxidized and that determination of M_n is accurate)

After 60 minutes of sodium hypochlorite oxidation at pH 7 the general degradation caused by the oxidation treatment is levelling off. Compared to the difference between "reference" and "30 minutes" further damage is less pronounced in this sample material. The carboxyl group content is still only slightly increased after deacidification treatment while changes in M_w and carbonyl group content are obvious (table 29).

Table 29. Overview of results for pulp samples with a carbonyl group content of 45.9 µmol/g (60 minutes of sodium hypochlorite treatment)

	M_w CCOA [kg/mol]	M_w FDAM [kg/mol]	C=O [µmol/g]	REG[1] [µmol/g]	COOH [µmol/g]
No treatment	185	151	45.9	36.3	22.7
Run A	155	131	41.4	32.1	23.0
Run B	161	130	41.2	33.7	23.8
Run C	153	132	43.2	30.1	23.5
Run D	153	136	43.8	33.1	23.5
Run E	159	138	43.1	35.5	excluded
Average [runs A – E]	156	133	42.5	32.9	23.4
SD [unit]	4	3	1.2		0.3
SD [%]	2	3	2.7		1.3

[1] REG: Reducing end groups calculated from M_n (please note: this value assumes that reducing ends are not further oxidized and that determination of M_n is accurate)

The degradation caused by the deacidification treatment can be followed by changes in molecular weight distribution between the treated and the untreated material (figure 71). Like in figure 72 after deacidification without previous oxidation treatment, some smaller molecular weight fractions get lost during treatment leading to a more narrow distribution. After the deacidification of oxidized sample material the content of carbonyl groups is decreased, while there is a slight increase in carboxyl group content.

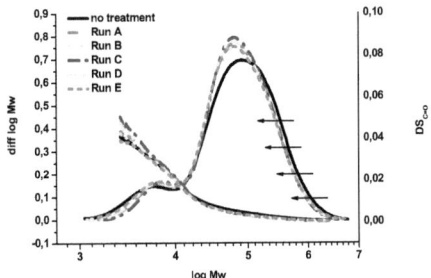

Figure 72. Molecular weight distribution and carbonyl group profile of sample sheets after 60 minutes of sodium hypochlorite treatment. Especially in the high MW region that is mainly responsible for strength properties a clear shift towards lower MW took place. This is a strong indicator for ongoing degradation caused by the deacidification treatment (black arrows).

To conclude the results obtained from pulp investigation it can be assumed that when dealing with pulps of more than 35 µmol/g carbonyl group content some degradation by β-elimination has to be expected, even though the carboxyl group content increased only slightly. Intensity and velocity of the process are driven by the amount and type of present carbonyl groups.

It should be noted that one drawback of the FDAM-labelling is its decreased sensitivity towards carboxyl groups other than uronic acids. Especially the hiding of some carboxylic groups due to the deacidification reagent that has not been removed completely during the acidic pre-treatment and formation of lactones and esters will lead to an underestimation of total amount of carboxyl group. This might be an explanation for the relatively low content of carboxyl groups detected in the present samples.

Further experiments on papers should enlighten the influence of additives and sizing agents on β-elimination. It is expected that they should slow down the degradative effect of alkaline treatments.

4.4.2 Sustainability of treatments

Usually mechanical testing is applied to test deacidification quality, but these test methods are not sensitive towards subtle changes. Two other parameters that are used to test the quality of deacidification treatment are pH, surface and extraction, and the determination of the alkaline reserve. Generally, also a homogeneous deposition of the alkaline reserve is desired. In routine work, usually only the surface pH is measured, because it is easy to perform and may be considered as non-destructive, even though it causes staining. It is well known, that pH does not correlate with the alkaline reserve [140], and occasionally the surface pH has already dropped below the desired values of > 7, but still an alkaline reserve can be detected [170].

Thus, it might be doubted, that a homogeneous surface pH will provide information about the homogeneity of the deacidification treatment. Until now, there is no knowledge available about correlation between the molecular conditions of paper and surface pH. In order to answer the question if an inhomogeneous distribution of surface pH after deacidification gives any information about the sustainability of the treatment, GPC analysis was applied to monitor the effect on molecular weight. This approach allows for the detection of subtle changes that might occur in relatively short periods of accelerated aging.

For the study, four different naturally aged papers (test papers "T", see page 48) in different conditions have been chosen based on their suitability for GPC analysis and required solubility in DMAc/ LiCl. A second criterion that was explicitly searched for in this study was an inhomogeneous deacidification reflected by a surface pH of less than pH 7 in at least one testing position. Surface pH and M_w were monitored on five different positions of the paper sheet before aging and treatment, after aging with deacidification and without deacidification (table 30).

All four analyzed papers had an initial surface pH with low standard deviation (equal or less than 0.2 pH units). After deacidification not all papers reached a homogeneously distributed surface pH as it was expected from their homogeneous initial surface pH. Within test papers T1 and T18 the observed surface pH values had a large deviation in the magnitude of 1 pH unit. Therefore, and because at least one value was below pH 7, they were considered as "unhomogeneously deacidified". Test papers T3 and T4 had a standard deviation of only 0.5 pH units, respectively 0.3 pH units. They are evaluated as "deacidified".

Table 30. Overview of results for deacidified and untreated test papers after three weeks of accelerated aging. Standard deviation was calculated from n = 5.

Sample	M_w initial [kg/mol]	M_w untreated, aged [kg/mol]	M_w treated, aged [kg/mol]	Surface pH untreated	Surface pH treated, aged
T1	110	95 ± 6	101 ± 5	4.2 ± 0.2	7.8 ± 1.1
T3	194	142 ± 7	158 ± 4	5.3 ± 0.0	8.1 ± 0.5
T4	319	228 ± 14	270 ± 12	4.1 ± 0.1	9.2 ± 0.3
T18	220	192 ± 7	180 ± 13	3.7 ± 0.1	6.7 ± 0.9

The initial M_w was very different in all of the tested papers right from the beginning. The observed range covers values from 110 to 320 kg/ mol. With regard to M_w, three different phenomena were observed upon aging and deacidification that will be further outlined.

The aging did not cause significant changes in the M_w, nor did treated samples behave different from untreated ones (figure 73). In this example surface pH was homogeneously before deacidification treatment, but after the treatment, in the middle of the sheet (position T1_5) the surface pH was not increased above pH 7. However, this had no negative implications on the M_w of position T1_5.

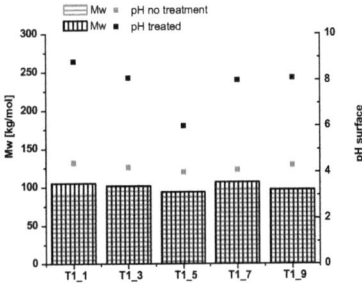

Figure 73. Comparison of M_w and surface pH with and without deacidification treatment of paper T1 after three weeks of accelerated aging. The initial M_w before accelerated aging was 110 kg/ mol. The aging did not cause significant decrease in M_w, nor do M_w values differ significantly between deacidified and untreated samples.

In another example, the aging caused significant decrease in M_w, but there is no significant difference between deacidified and untreated samples (figure 74). The significance of these results was mainly disturbed by sub sample T18_3 (upper right corner) when both results, for the untreated and the treated sample, had lower respectively higher values than the other four sub samples. For surface pH the same uneven distribution towards the centre parts of the sheet as for test paper T1 was observed.

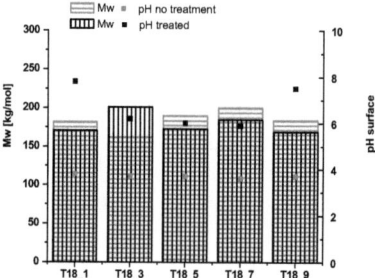

Figure 74. Comparison of M_w and surface pH with and without deacidification treatment of paper T18 after three weeks of accelerated aging. The initial Mw before accelerated aging was 220 kg/ mol. The aging caused significant decrease in M_w, but there is no significant difference between deacidified and untreated samples.

In the last two samples in this test series, the aging caused significant decrease in M_w, but the treated samples are significantly better preserved (figure 75). This tendency is more pronounced in test paper T4 than in T3. Both papers exhibit additionally a relatively homogeneously distributed surface pH, indicating an even deacidification of the sheet.

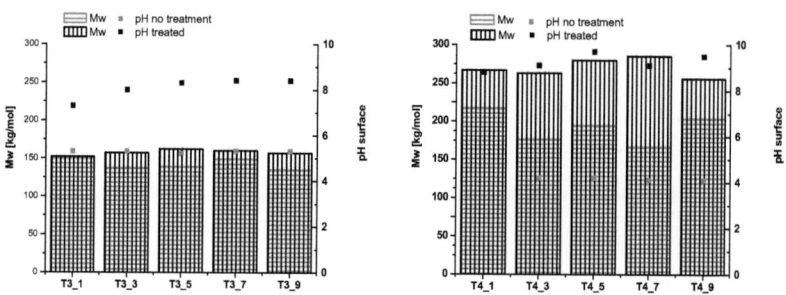

Figure 75. Comparison of M_w and surface pH with and without deacidification treatment of paper T3 (left, initial M_w before accelerated aging of 194 kg/ mol) and paper T4 (right, initial M_w before accelerated aging of 320 kg/ mol) after three weeks of accelerated aging. The aging caused significant decrease in M_w, but treated samples are significantly better preserved.

The small quantity of samples analyzed for this study does not allow for an overall estimation of the influence of surface pH on sustainability of deacidification treatments. Some tentative conclusions might be deduced from the four present samples:

- When a paper with a low initial M_w is aged (T1), further changes do not readily occur. As there are no information available about the condition of the paper when it was produced it is difficult to assess its present condition. Nevertheless, as no additional decrease in M_w occurs and as no differentiation between treated and untreated sample material is observed it can be assumed, that a levelling-off degree of M_w has been reached in this sample. Deacidification of this type of sample material will not lead to any added benefit, because the degradation is too advanced already.

- The opposite is true for a paper with high initial M_w (T4 and T3). Even though the paper suffers from degradation in both cases, deacidified and not treated, the positive effect of deacidification can be clearly established.
- Even though there is a correlation between significantly better preserved M_w and a homogeneous distribution of surface pH (T3 and T4), it cannot be concluded that an inhomogeneous distribution of surface pH will necessarily exert detrimental influence on the paper, simply because in the samples analyzed there was no significant difference in the M_w.
- On the other hand it is quite obvious that the initial pH will not predict the behaviour towards deacidification treatment, neither in terms of homogeneous distribution of the initial value nor in terms of the absolute initial value. Taking standard deviation into account, both papers, T1 and T4 have the same initial surface pH, but they behave differently upon deacidification and accelerated aging. The same observation is made for samples with a low standard deviation as found for papers T4 and T18.

4.4.3 Summary

The main application of mass deacidification is needed for papers produced between 1850 and 1980. In these papers cellulose is only one among several other components as ground wood pulps have been mainly used for paper production. The success of mass deacidification treatments can be monitored using GPC on a molecular level which makes it very attractive as a means of quality control; because property changes can be linked directly do molecular parameters, even before mechanical consequences can be measured.

When deacidification is performed on historic papers, this paper has not only suffered from hydrolytic degradation caused by acids but in the course of time also oxidized functionalities have been introduced. Sustainability of mass deacidification treatments have to be analyzed to prove the benefit of the treatment towards possible disadvantages like β-elimination. Experiments based on pulp in various stadiums of oxidation show that when more than 35 µmol/g carbonyl group content is present some degradation by β-elimination has to be expected.

A detailed look at the influence of deacidification within a sheet of naturally aged paper mainly reveals that treated papers generally are more resistant towards accelerated aging than non-treated papers. For some papers analyzed this effect is significant, in others rather a trend is depicted. It shows that initial surface pH cannot predict the behaviour towards mass deacidification and further aging. The same is true for surface pH after treatment, even though significantly better preserved papers tend to have a homogeneous distribution of surface pH in the alkaline region as well.

The results of the sustainability study of mass deacidification indicate that rather than disadvantages caused by degradation via β-elimination in the alkaline region the protection of cellulose due to neutralized acids and alkaline reserve seems to be of greater importance. These findings are underlined by the fact that historic papers do not consist of cellulose and some hemicelluloses like pulp hand sheets, but are complex materials with many buffering elements that prevent cellulose from being attacked by β-elimination due to strong alkalinity after the treatment.

General condition rating of historic papers could profit from this type of studies. Classes of damage or preservation could be established by defining threshold values. The analysis of the influence of β-elimination might serve as an example for this approach. It was found that a certain amount of carbonyl groups has to be exceeded before β-elimination can be detected at all. According to this model several other treatments could be studied.

4.5 Improvement of cellulose solubility

When the question of mass deacidification is addressed, usually papers produced between 1850 and 1980 have to be analyzed. One characteristic of these papers is a high content of ground wood. Lignin and hemicelluloses, contained in large amounts in ground wood pulps, exert a negative impact on pulp solubility. The dissolution in DMAc/ LiCl 9 % prior to GPC analysis is rendered difficult.

The major obstacle for softwood Kraft pulp dissolution was found to be stiffness, *i.e.* low swellability, of the fibre in high kappa number pulps [51]. This is certainly also true for pulps containing large amounts of ground wood. Nevertheless, isolated lignin is easily soluble in the DMAc/ LiCl-system and even fully bleached softwood Kraft pulps could not be completely dissolved [51]. When analyzing undissolved residues of softwood Kraft pulp it was found that the content of lignin and mannans is raised towards native pulp samples. It was suggested that associations and/ or bonds might exist between the two components [171]. Even when raw materials for paper production have been pulped, hemicelluloses are known to possess gelling properties that hinder cellulose accessibility. Hemicelluloses partly dissolve upon Kraft cooking conditions and, in a later stage of the cooking process, re-precipitate on the cellulose fibres [172].

To circumvent the solubility problems, enhanced cellulose solubility was studied to make more papers, especially those containing important amounts of ground wood, available for analysis. Closely related to that topic is the question in how far these solubility enhancement procedures will damage cellulose or negatively influence the fluorescence labels used for cellulose analysis.

4.5.1 Study of cellulose degradation in derivatization systems

In a detailed study the presence of dimethylsulfonium ions and the derived methyl(methylene) sulfonium ylide in carbanilation mixtures consisting of DMSO/ Ph-NCO/ cellulose was proved [173]. The presence of a trapping agent and its subsequent extraction into n-hexane provided a mixture of non-reacted trap and a main product, the corresponding cyclopropane derivative that was identified by several analytical techniques. A blank experiment was performed to confirm the absence of cyclopropanated product in pyridine/ DMSO carbanilation system, with DMSO alone and with Ph-NCO alone.

The next step was the demonstration of the direct chemical interaction of sulfoxide and cellulose beyond a general oxidizing effect [173]. As DMSO-containing carbanilation mixtures follow the basic chemistry of the Moffatt and Swern oxidation systems with regard to oxidation behaviour, they should also exhibit the characteristic behaviour of such systems in terms of side reactions. Methylthiomethyl ethers should be formed as by-products according to a Pummerer-type rearrangement, involving cellulosic hydroxyl groups (figure 76). The chemical interaction of sulfoxide solvents with cellulose does not only consist in an oxidative effect as the major pathway, but also in a derivatizing effect as a side reaction.

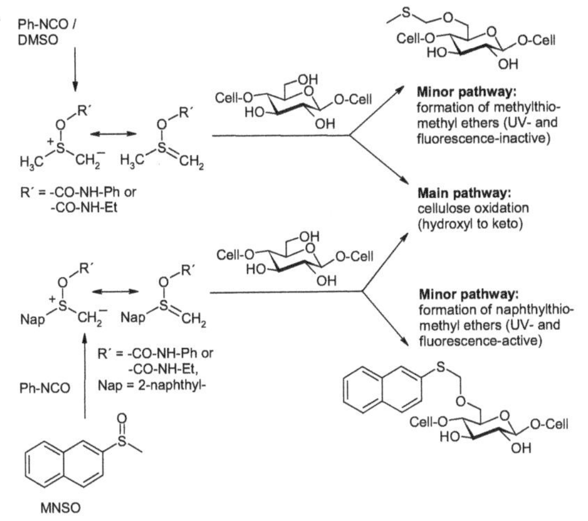

Figure 76. Confirmation of the chemical reaction between cellulose and sulfoxide-derived intermediates: reaction mechanism and derivatives. Cellulosic carbamoyl substituents are not shown.

In the case of DMSO as the source, such (thio)ether moieties are quite difficult to detect on cellulose due to their very low concentration which is far beyond a reliable detection by elemental analysis. In order to circumvent this obstacle for the study of the oxidizing action of isocyanate derivatisation on cellulose, methyl-(2-naphthyl) sulphoxide (MNSO) was synthesized and used as a molecular probe. By replacing one methyl group of DMSO with a 2-naphthyl residue this molecule is able to form sulphoxonium species and the corresponding ylide similar to the DMSO-derived ylide. Due to its photoactive naphthyl moiety UV or fluorescence detection is achieved. The reactivity difference between DMSO and MNSO with regard to ylide formation is negligible. After gel chromatography analysis of derivatized cellulose the attached sulphur-containing groups originating from side reactions with the sulphoxide species can therefore be reported. Since phenyl isocyanate is UV-detectable as well, ethyl isocyanate was used being an UV- and fluorescence-silent derivatizing reagent.

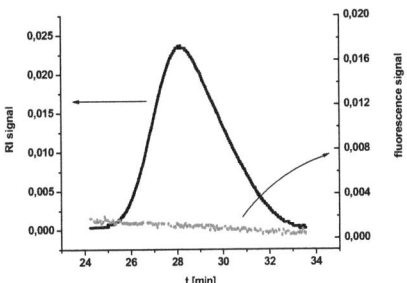

Figure 77. Gel permeation chromatogram of cotton linters pulp derivatized with ethyl isocyanate in pyridine. No fluorescence signal is detectable.

The reaction of cellulose with ethyl isocyanate in pyridine or DMSO caused no UV/ fluorescence to appear (Figure 77). When MNSO was added into the carbanilation mixture to replace 10 mol% (0.1 eq.) of the solvent DMSO, cellulose ethylcarbanilates exhibited considerable UV/ fluorescence activity upon GPC analysis. Gel permeation chromatography (GPC) confirmed that derivatization of cellulose in DMSO introduces thiomethyl ether moieties in very small amounts. This was demonstrated for two different pulps, cotton linters (figure 78) and beech sulphite dissolving pulp. The derivatized cellulose exhibited a distinct fluorescence signal, which followed the molecular weight distribution, indicating a uniform introduction without significant discrimination effects. This proved that the naphthylthiomethyl residues were covalently bound to the cellulose backbone. Simple adsorption would have caused fluorescence activity in the exclusion peak which was not present.

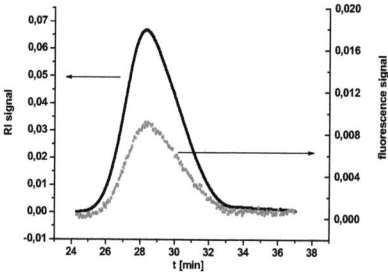

Figure 78. Size-exclusion chromatogram of cotton linters derivatized with ethyl isocyanate in DMSO containing 10 mol% (relative to DMSO) MNSO.

In order to get a rough estimation of the degree of substitution (DS), a cellulose sample was derivatized with 0.01% 2-naphthyl isocyanate (relative to the available hydroxyl groups) and subsequently carbanilated with ethyl isocyanate (figure 79). It can be roughly calculated that the DS of oxidized groups introduced by the carbanilation in DMSO is about 40 – 100 times larger than the amount of Pummerer products formed [173]. Carbanilation in sulfoxide-based solvents will introduce keto groups in the range between 2 per 1000 and 2 per 100 anhydroglucose units (AGU).

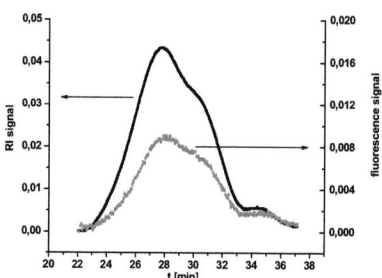

Figure 79. Size-exclusion chromatogram of cellulose (beech sulphite pulp) derivatized with 2-naphthyl isocyanate (0.1 % relative to cellulosic OH) in pyridine followed by reaction with ethyl isocyanate. The fluorescence signal indicates presence of 2-naphthylcarbamoyl substituents over the whole molecular weight range.

Carbanilation of celluloses by isocyanates in DMSO will cause degradation of cellulose and act as "derivatizing agent" according to a side reaction mechanism. In DMSO-based mixtures, cellulose carbanilation will therefore negatively affect cellulose by oxidation. These sulphur containing species cause also the oxidative effect of the DMSO/ isocyanate carbanilation medium on cellulose. This oxidation introduces carbonyl functionalities along the cellulose chain that constitute spots of pronounced chemical instability. Under basic conditions according to β-elimination processes, subsequent cleavage will readily occur. Short reaction times and low derivatization temperatures will minimize the negative effects, but oxidation cannot be completely avoided. Thus, when applications aim at molecular weight determinations or molecular weight distribution, DMSO as the solvent or as a solvent component should not be used.

4.5.2 Study of improvement of cellulose solubility

Cellulose solvent systems are generally divided into two main classes:
- Non-derivatizing methods - cellulose is directly dissolved (NMMO, DMAc/ LiCl, metal complexes)
- Derivatizing methods - cellulose is chemically transformed into its derivative before dissolution (cellulose carbamates via reaction with isocyanates)

Especially the DMAc/ LiCl system, a non-derivatizing method, is a very good solvent for pure cellulose. Nevertheless, when it comes to pulps and papers containing large amounts of lignin and hemicelluloses the dissolution is complicated, in particular for pulps with high molecular weight. Other obstacles for successful dissolution of paper are the content of mineral fillers and sizing agents. All these limitations hold true for papers in the context of conservation: historic papers, especially from the 19th and 20th century, contain hemicelluloses, lignin, sizing and mineral fillers.

As described in the Materials and Methods section (see page 47), *NOVO* paper consists of both, chemical and ground wood pulp plus mineral fillers and alum rosin sizing. It has therefore been chosen as a model to test the capability of the various approaches to improve the solution of cellulose. In a preliminary test, it was found that *NOVO* paper is neither soluble in the standard dissolution system DMAc/ LiCl 9 % (non-derivatizing solvent system) nor after derivatisation in pyridine/ phenyl isocyanate (table 31).

Table 31. Two standard procedures to dissolve NOVO paper, DMAc/ LiCl 9 % and carbanilation in pyridine by phenyl isocyanate. All samples were disintegrated in a mixer.

No.	Pre-treatment	Solvent exchange	Activation	Results
1	None	Ethanol and DMAc over night	DMAc/ LiCl 9 %	No solution, turbid, no increase in viscosity, fibres
2	None	DMAc	Freeze drying, pyridine + phenyl isocyanate	Partly soluble, dissolved portion yields white waxy substance

Immediately after the direct dissolution approach with DMAc/ LiCl 9 % no visible sign of solution was detected. Only after more than one year, some swelling was observed under the microscope and viscosity increased. The silhouette of the single fibres became blurred. Cross sections were still found to be intact. The derivatizing method was able to dissolve at least some part of the paper. After precipitation in water that serves to regenerate the dissolved cellulose, a white waxy substance was obtained that did not resemble anything that was obtained after other carbanilation experiments. Thus, it was concluded that cellulose was not dissolved but some other component.

In order to improve cellulose solubility several attempts based on the DMAc/ LiCl system have been made. They are divided into extended standard approaches, mixed systems and alternative approaches, and will be explained more in detail in the following section. The state of dissolution in all approaches was evaluated visually by the unaided eye, unless stated differently.

Extended standard approaches

The standard approach based on DMAc/ LiCl of pulp or paper dissolution is done by a sequence of mechanical disintegration, solvent exchange, activation and dissolution. This approach has been extended using steam explosion, reduction or extraction steps before dissolving (table 32). One important aspect for the dissolution of cellulose is the accessibility of the cellulose for the dissolving agent. As aqueous disintegration of the sample in a mixer for about one minute is obviously not enough to sufficiently increase accessibility, several other approaches were tried.

Examples no. 1a and 2a are based on previous Soxhlett extraction using acetone and dichloromethane to remove parts of the sizing agent that additionally impede access to the fibre. The extraction was completed by additional activation using DMAc over night. The sizing agent of *NOVO* paper is alum rosin system. Rosin consists mainly of abietic acid, and combines with caustic alkalis to form salts (rosinates or pinates) that are known as rosin soaps.

Another idea was followed in examples no. 3a and 4a. These approaches aimed at the improved activation of cellulose via freeze drying and steam explosion to open the cellulose structure mechanically. Both approaches have been used successfully to improve accessibility. Freeze drying was suggested to improve dissolution of cellulose [50, 174], while steam explosion was studied in the context of improving enzymatic activity on pulps [175, 176].

Examples no. 5a and 6a are based on a chemical alteration of the pulp via reduction (further descriptions on page 51). Both, $NaBH_4$ and borane tert-butylamine (TBAB) partly reduce carbonyl groups to the corresponding hydroxyl groups and are therefore reported to stabilize pulp in bleaching sequences [177]. It was found only recently, that high contents of carbonyl groups contribute to cross-linking reactions via hemiacetal linkages, their reduction to hydroxyl groups should therefore improve accessibility [178, 179]. Even though the amount of carbonyl groups in *NOVO* paper is unknown it was considered worth a try to perform a reduction.

Table 32. Extended standard approaches to improve NOVO solubility. All samples have been disintegrated in a mixer, the solvent was always DMAc/ LiCl 9 %.

No.	Pre-treatment	Solvent exchange	Activation	Results
1a	Extraction with acetone	none	DMAc over night	No solution, turbid, no increase in viscosity, fibres
2a	Extraction with dichloromethane	none	DMAc over night	No solution, turbid, no increase in viscosity, fibres
3a	None	Ethanol, DMAc	Freeze drying	No solution, turbid, no increase in viscosity, fibres
4a	None	Ethanol, DMAc	CO_2 steam explosion[12] (40°C, 250 bar, 1h)	No solution, turbid, no increase in viscosity, fibres
5a	Reduction with $NaBH_4$ (24h)	Ethanol	DMAc over night	No solution, bleaching, slight increase in viscosity, fibres
6a	Reduction with TBAB (24h)	Ethanol	DMAc over night	No solution, turbid, no increase in viscosity, fibres, bleaching

When using the standard approach with extended procedures to improve accessibility like extraction, freeze drying, steam explosion and reduction, no significant improvement in solubility can be observed. There is some viscosity increase for sample 5a ($NaBH_4$) that was interpreted as improved swellability, but no full dissolution was achieved. Nevertheless, the application of an alkaline substance like $NaBH_4$ on oxidized pulp samples, β-elimination leading to chain scission will be faster than reduction (unpublished results[13]). Therefore, when solubility is improved some degradation of cellulose is expected to contribute to the improvement, too. Microscopic analysis of samples 5a and 6a detected fibres that were split into single compartments while the connection between them has disappeared.

Interestingly, the viscosity of sample 2a has increased after prolonged dissolution time, i.e. after more than one year. Observation by microscope revealed some swelling of fibres, mainly exhibiting a blurred silhouette. Cross sections of fibres were not affected by the starting swelling process.

Mixed systems

When applying mixed systems, the direct dissolution power of DMAc/ LiCl is exploited and supported by an extra derivatization step using isocyanates (table 33). The idea is to partly derivatize OH-groups in the cellulose and thereby reduce the ability to form hydrogen bonds. Even though the combination of derivatizing agent and direct solution has already been proven to successfully improve cellulose solubility in softwood kraft pulps [49], the high quantity of ground wood in *NOVO* still poses a difficult task for the mixed systems approach.

Examples no. 1b to 4b follow the guidelines for improved cellulose solubility established at the *STFI* [180]. The *STFI* standard method was designed for routine work and requires only relatively short solvent exchange and activation procedures. It is shortly described in example 1b. In examples 2b to 4b, this standard approach was slightly modified in terms of prolonged solvent exchange to improve cellulose accessibility.

In example 5b the standard method was translated into the standard approach followed at *BOKU* as already described in table 30, example no. 1. The main differences between the

---
[12] Steam explosion experiments were performed by DI Emmerich Haimer, Department für Materialwissenschaften und Prozesstechnik at BOKU Vienna
[13] Own experiments

two protocols are the mechanical disintegration prior to solvent exchange and a higher concentration of lithium chloride (9 % instead of 8 %) in the *BOKU* method.

In examples 6b to 8b the different dissolution procedures were separated, *i.e.* first derivatization was accomplished by the addition of isocyanate, and only later lithium chloride was added as direct solvent. In examples 7b and 8b several extraction steps were performed prior to solvent exchange to additionally ease penetration properties of the solvent.

Table 33. Mixed approach to improve the solubility of NOVO paper. All samples with an * have been disintegrated in a mixer before pre-treatment.

No.	Pre-treatment	Solvent exchange	Activation	Solvent	Results
1b	None	Water (1h)	3x DMAc	DMAc/ LiCl 8% + EIC	No solution, turbid, no increase in viscosity, fibres
2b	None	water (over week end)	3x DMAc	DMAc/ LiCl 8% + EIC	No solution, turbid, no increase in viscosity, fibres
3b	None	Acetone (over week end)	3x DMAc	DMAc/ LiCl 8% + EIC	No solution, turbid, no increase in viscosity, fibres
4b	None	Ethanol (over week end)	3x DMAc	DMAc/ LiCl 8% + EIC	No solution, turbid, no increase in viscosity, fibres
5b	None*	Ethanol	DMAc	DMAc/ LiCl 9% + EIC	No solution, turbid, no increase in viscosity, fibres
6b	None*	Ethanol and 2x DMAc over night	DMAc + EIC (5 days)	LiCl 9%	Swelling, slight increase in viscosity, turbid liquid, fibres visible
7b	Extraction with Acetone*	DMAc over night	DMAc + EIC (5 days)	LiCl 9%	No solution, turbid, no increase in viscosity, fibres
8b	Extraction with di-chloro methane*	DMAc over night	DMAc + EIC (5 days)	LiCl 9%	No solution, turbid, no increase in viscosity, fibres

While for other samples the mixed approach combining derivatzation and DMAc/ LiCl proved to work well (results not presented in this work), the dissolution of *NOVO* was not achieved. One possible explanation is that the lignin contained in ground wood is by far higher than the pulp samples used for the development of this method. *NOVO* contains between 50 to 65 % of ground wood. Based on the fact that softwood ground wood generally contains between 25 to 30 % of lignin, the *NOVO* paper will contain between 12 to 20 % of lignin, while the samples used for the development of the improved solubility method at *STFI* had a maximum lignin content of 8 %.

For example 6b an increased viscosity was found that is usually a sign of beginning swelling of cellulose. Nevertheless, it remained unclear whether this increased viscosity was due to starting dissolution or rather dissolution problems encountered by the late addition of lithium chloride.

Again, the additional extraction step did not improve solubility. Probably the sizing is not removed at all. One theory might be that sizing will prevent cellulose solution as efficiently as high lignin contents.

Alternative approaches

Alternative approaches are based on different solvents instead of DMAc, even though still LiCl was used (table 34). Like DMAc, alternative solvents are polar and aprotic. Especially DMSO was found to have one of the most efficient swelling capacities for cellulose [4]. Therefore, it forms part of several cellulose dissolving agents as well as described in chapter 4.5.1. DMI was presented as a solvent system that improves cellulose solubility [181].

Table 34. Alternative approaches to improve NOVO solubility. All samples have been disintegrated in a mixer before pre-treatment.

No.	Pre-treatment	Solvent exchange	Activation	Solvent	Results
1c	none	Pyridine	Pyridine + EIC (5 days)	LiCl 9%	No solution, turbid, no increase in viscosity, fibres, yellowing
2c	none	DMSO	DMSO + EIC (5 days)	LiCl 9%	No solution, turbid, no increase in viscosity, fibres, yellowing
3c	none	Ethanol	DMI (over night)	DMI/ LiCl 9%	Strongly increased viscosity, but no complete solution, turbid, fibres

Best results have been achieved when replacing the DMAc/ LiCl system by DMI/ LiCl. Under the microscope the swelling of single fibres is observed, characterized by a blurred silhouette. Nevertheless, even though viscosity was considerably increased and indications for dissolution can be found, no complete solution of the *NOVO* sample has been achieved. The state of swelling remained constant over an extended period of time, no further improvements were observed after more than one year.

The most unexpected observation was the strong yellowing that occurred when replacing DMAc by DMSO and pyridine. It was not further investigated what caused the yellowing, but as all other dissolution systems described in this section did not cause yellowing it is regarded as negative, probably indicating oxidation or formation of by-products that might negatively influence cellulose properties.

4.5.3 Study of the influence of EIC on pulp

Several attempts to increase cellulose solubility described in chapter 4.5.1 and 4.5.2 are based on the use of ethyl isocyanate (EIC) as derivatizing agent. It is reported that the degree of substitution in this system is around 2 [180]. That means out of three possible hydroxyl groups available at the cellulose molecule, one or two will be replaced by ethyl isocyanate (figure 80). Consequently, a maximum increase of molecular weight by a factor of 1.87 taking two additional isocyanate groups into account should be observed.

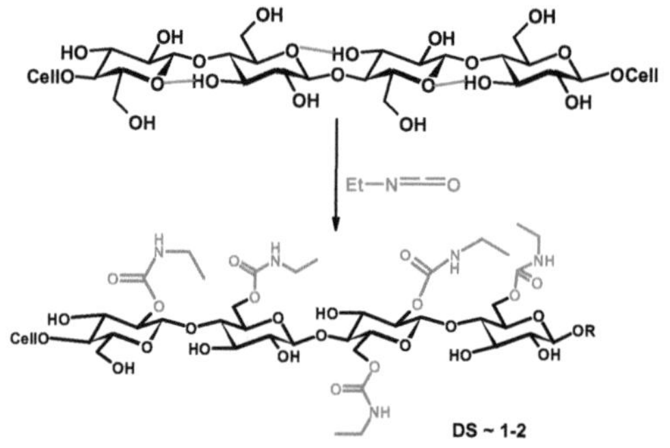

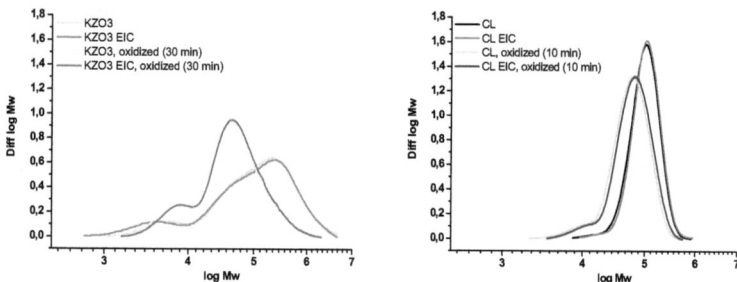

Figure 80. The cellulose carbamate after derivatization by ethyl isocyanate is expected to have a degree of substitution DS ~ 1-2.

This increase in molecular weight should be reflected in the molecular weight distribution: the hydrodynamic value of such a substituted molecule is bigger and therefore, derivatized molecules should elute faster. Nevertheless as can be seen in figure 81, no pronounced shift in the MWD as expected took place. For the derivatization experiments two native (see page 44 - 45) and two oxidized pulps (see page 49) were used in order to study the influence of decreased availability of hydroxyl groups for derivatization. No influence, except for general degradation of the pulp by the oxidation procedure, was found.

Figure 81. Different pulps without and after EIC-derivatization. No significant change in molecular weight distribution is observed. Left: Native beech sulphite dissolving pulp (BKZO) and after 30 minutes oxidation. Right: Native cotton linters (CL) and after 10 minutes oxidation.

This leads to the conclusion that either dn/dc has changed because of the derivatization procedure or the degree of substitution is lower than expected. When calculating the expected M_w of pulp substituted by two ethyl isocyanate moieties, thus an increase of the M_w by a factor of 1.87, it was found that the dn/dc will have to be changed considerably to reach these values. It is not very likely that such a small substitute like ethyl isocyanate will cause such severe modifications in dn/dc. Therefore the later theory is favoured. That would imply that MW determination by laser light scattering of cellulose derivatized by ethyl isocyanate could be performed without modifications of the dn/dc.

4.5.4 Isocyanates in combination with fluorescence labelling

Next to the question in how far cellulose solubility can be improved, it was also of interest to find out if the improved solution system is still compatible with fluorescence labelling. In the following chapter the influence of EIC as derivatizing agent was tested on well characterized standard pulps after CCOA and FDAM-labelling. One main obstacle in a reliable detection of fluorescence was expected to be the addition of methanol to quench EIC after accomplished derivatization, because it is also known to quench fluorescence. Next to standard pulps with different amounts of carbonyl and carboxyl groups, additional analysis was performed on oxidized pulps.

EIC and CCOA-labelling

It was found that in combination with EIC derivatization of standard pulps the fluorescence intensity after CCOA-labelling was considerably decreased (figure 82). Only less than 10 % of the fluorescence that was measured without addition of EIC was detected (table 35).

Table 35. Overview of the effect of the derivatising agent ethyl isocyanate and the quenching agent methanol on detected carbonyl group content.

Sample name	C=O [µmol/g]	C=O$_{EIC}$ [µmol/g]	Δ C=O$_{EIC}$ [µmol/g]	C=O$_{MeOH}$ [µmol/g]	Δ C=O$_{MeOH}$ [µmol/g]
st1	0.8			0.5	0.3 (62.5 %)
st3	5.1	0.2	4.9 (3.9 %)	4.6	0.5 (90.2 %)
st5	11.1	0.7	10.4 (6.3 %)	11.7	0.6 (105.4 %)
st7	15.7	1.3	14.4 (8.3 %)	16.0	0.3 (101.9 %)
st8	23.1	2.0	21.1 (8.7 %)	22.5	0.6 (97.4 %)
st6	30.7	2.5	28.2 (8.1 %)	29.1	1.6 (94.8 %)

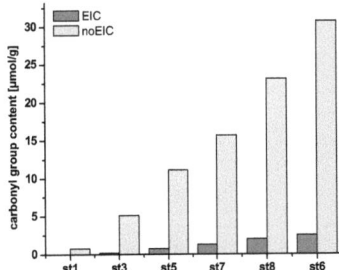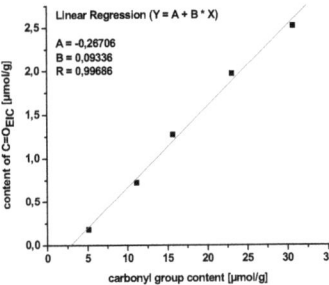

Figure 82. Left: Carbonyl group content of EIC treated and non-treated standard pulps (samples contain methanol to inactivate EIC). Right: Carbonyl group content of methanol treated and non-treated standard pulps

As methanol is known to be an efficient quenching agent for fluorescence the experiment was repeated with addition of methanol like in the EIC-experiment, but without addition of EIC itself. It was found that fluorescence intensity was not significantly disturbed by the addition of methanol only (figure 83). Just data of standards 1 and 3 slightly deviated, maybe because the total amount of carbonyl groups is relatively low (table 34). Values above 10 µmol/g were determined in the range of expected errors.

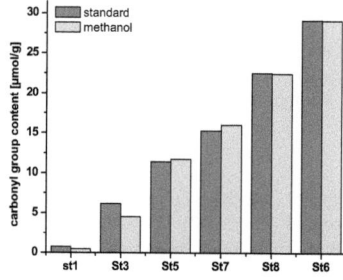 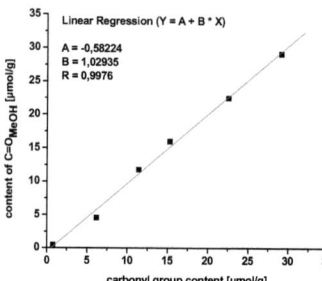

Figure 83. Left: Carbonyl group content of EIC treated and non-treated standard pulps (samples contain methanol to inactivate EIC). Right: Correlation of carbonyl group content of EIC treated and non-treated standard pulps

EIC and FDAM-labelling

After FDAM-labelling the pulps show a different behaviour than after CCOA-labelling. The detected amount of carboxyl groups decreased only slightly in combination with EIC. Like for carbonyl groups, amounts lower than 10 µmol/ g exhibited lower response to fluorescence detection while the detection of higher overall values was hardly influenced (table 36).

Table 36. Overview of the effect of the derivatising agent ethyl isocyanate on detected carboxyl group content.

Sample name	COOH [µmol/g]	COOH$_{EIC}$ [µmol/g]	Δ COOH$_{EIC}$ [µmol/g]
St1	8.9	7.3	1.6 (82.0 %)
St2	11.8	10.7	1.1 (90.7 %)
St3	17.5	n.d.	n.d.
St4	20.6	18.6	2.0 (90.3 %)
St5	20.5	n.d.	n.d.
St6	22.6	n.d.	n.d.
St7	32.2	30.3	1.9 (94.1 %)

n.d.: values not determined

Additional experiments were performed on oxidized pulps. Cotton linters was oxidized by TEMPO for 3 to 10 minutes and compared to the starting material after derivatization using EIC (figure 84). TEMPO oxidation is known to induce selective oxidation of C6 at the cellulose molecule [182]. Increasing the rate of oxidation does not negatively influence the detection of fluorescence after FDAM-labelling.

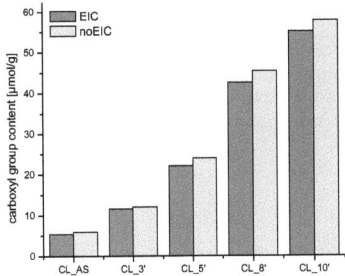 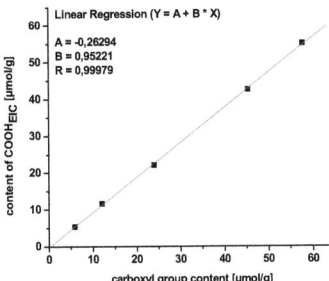

Figure 84. Influence of EIC on the detection of carboxyl group content in cotton linters after increasing oxidation. Left: Cotton linters after different oxidation steps to produce increased amount of carboxyl groups with and without EIC. Right: Correlation of carboxyl group yield between standard labelling procedure and labelling procedure including EIC.

The same type of experiment was repeated on dissolving beech sulphite pulp (BKZO). This pulp contains hemicelluloses and is therefore a more complex example than cotton linters that consists to almost 100 % out of cellulose. The starting material was compared to pulp oxidized by TEMPO for 3.5 to 30 minutes. The behaviour is comparable to the two previously obtained results: there is no negative influence of increased oxidized functionalities towards the addition of EIC after FDAM-labelling (figure 85).

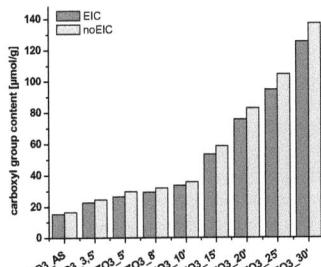 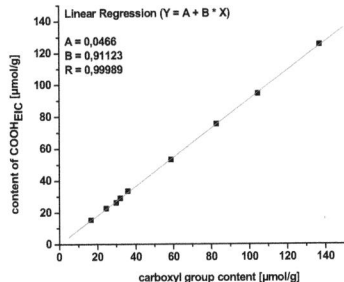

Figure 85. Influence of EIC on the detection of carboxyl group content. Left: Cotton linters after different oxidation steps to produce increased amount of carboxyl groups. Right: KZO3 after different oxidation steps to introduce increased amount of carboxyl groups.

The two labels used for the fluorescence labelling approach followed by GPC-analysis react differently towards the derivatizing agent EIC. While CCOA does not tolerate the application of EIC and reacts with considerably lowered fluorescence yield, FDAM still works in a satisfying way. It was shown that the decreased fluorescence intensity is not due to methanol. The addition of methanol has a very low influence on fluorescence.

However, a linear correlation was established for all examples investigated, CCOA and FDAM standards, and oxidized pulps. The correlation coefficient is well above 0.99 in all cases analyzed. It was also shown that correlation between values obtained with and without addition of EIC is the better the higher the absolute amount of oxidized functionalities is.

Therefore, when sensitivity is not a problem, CCOA and especially FDAM may still be used in combination with EIC. Nonetheless the error of measurement will be increased.

The decreased availability of C6 after TEMPO oxidation does not have any influence on the derivatization. This supports the theory that the degree of substitution of hydroxyl groups by ethyl isocyanate is considerably lower than 2 because the two competing reactions do not seem to interfere with each other even when carboxyl group content exceeds 100 µmol/g.

4.5.5 Summary

In this chapter the influence of derivatizing systems on cellulose was studied, namely isocyanates in an organic solvent.

The carbamation of celluloses by isocyanates in DMSO will cause degradation of cellulose and act as "derivatizing agent" according to a side reaction mechanism. This negative effect on cellulose is oxidation that introduces carbonyl functionalities along the cellulose chain. Under basic conditions according to β-elimination processes, subsequent cleavage will readily occur. It is recommended thus, not to use DMSO in the derivatizing system when determination of molecular weight or molecular weight distribution is aimed at.

One of the papers suggested for quality control of mass deacidification (*NOVO*) is basically not suitable for GPC measurements. It does not dissolve in the standard dissolution system DMAc/ LiCl (9%) and in a pyridine/ ethyl isocyanate derivatization reaction. Best results immediately after preparation of the samples have been achieved after previous reduction with NaBH$_4$, when adding lithium chloride after the derivatizing procedure and when replacing the DMAc/ LiCl system by DMI/ LiCl. Nevertheless, even though viscosity was clearly increased, no complete solution of the *NOVO* sample has been achieved. Additionally, it remained unclear, especially after the later addition of lithium chloride and after replacement of DMAc by DMI, if a real improvement in solubility was achieved or if rather decreased solubility of the lithium chloride or changed viscosity parameters of the solvent caused the higher viscosity observed. Further microscopic analysis after one year supports the theory that via DMI/ LiCl best and fastest part-solution was achieved, because swelling of the fibres was clearly established. Nevertheless, the same state of viscosity and fibre swelling was obtained after the standard dissolution method and after previous extraction with dichloromethane, but it obviously took a lot longer to obtain this result. Reduction with NaBH$_4$ leads to a different type of dissolution, some fibre sections dissolve, while others remain intact. In case of *NOVO* paper the application of EIC cannot be recommended. Its use is not justified because better results are obtained with less toxic substances.

It was also concluded that the degree of substitution after EIC-derivatization is lower than formerly reported. Changes of dn/dc are not considered to be very likely, mainly because the labelling of increased amount of carboxyl groups due to TEMPO oxidation does not interfere with derivatization.

Further analysis showed that the fluorescent labels react differently towards the derivatizing agent EIC. While CCOA does not tolerate the application of EIC and reacts with considerably lowered fluorescence yield, FDAM-labelling works still satisfying. It was shown that the decreased fluorescence intensity is not due to methanol.

However, a linear correlation was established for all examples investigated, CCOA and FDAM standards, and oxidized pulps. The correlation coefficient is well above 0.99 in all cases analyzed. It was also shown that correlation between values obtained with and without addition of EIC is the better the higher the absolute amount of oxidized functionalities is. Therefore, when sensitivity is not a problem, CCOA and especially FDAM may still be used in combination with EIC. Nonetheless the error of measurement will be increased.

4.6 Development of a non-destructive approach

For the development of a non-destructive approach to assess the condition of historic papers, near infrared spectroscopy was chosen. Band assignment of NIR-spectra (Figure 86 left) is rather complicated due to many broad and overlapping bands corresponding to overtones and combinations of fundamental vibrations appearing in the mid infrared region. Even though pulp hand sheets and historic rag papers are completely different materials, their spectra look very much the same. The drawback of lacking possibility to assign peaks to certain functional groups is compensated for by multivariate data evaluation, a process capable of finding hidden correlations. As the evaluation of spectra is based on a statistical approach, variations in sample appearance that would adversely affect mid infrared spectra are overruled.

This non-destructive approach aims at predicting the condition of historical papers. Before models of historic papers were attempted, a feasibility study on pulp hand sheets was performed, mainly because it is a less complex material without fillers, sizing agents and traces of natural aging and use. Models for pulp hand sheets might find their application in pulp industry. The first sub-chapter is dedicated to the development of prediction models based on pulp hand sheets (see page 45), and a second sub-chapter deals with the modelling of historic rag papers (see page 48).

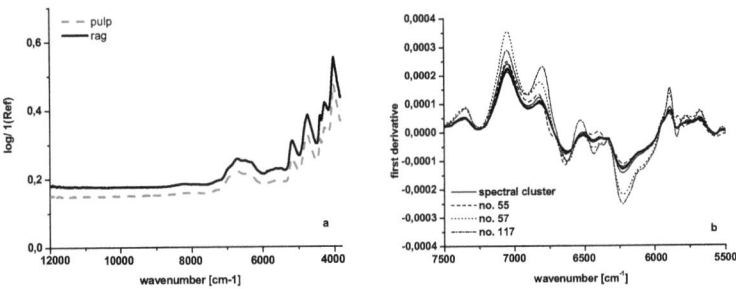

Figure 86. Left: NIR-spectra of pulp hand sheet (dashed line) and rag paper (solid line). Right: First derivative of pulp NIR-spectra (detail) showing visually deviating pulp samples no. 55, no. 57 and no. 117 that have been excluded from multivariate calibration.

Pre-processing of spectral data is useful to account for differences within the set of sample papers, especially for historic rag papers that are obviously different in colour and thickness. Derivatives, multiplicative scattering correction, vector normalization and straight line subtraction, or combinations thereof, were found to be appropriate for the present problem. Before and after pre-processing, the spectra have been visually examined for unusual features. In figure 86 right this selection procedure is demonstrated for pulp hand sheets. Three spectra (no. 55, 57, and 117) of pulp hand sheets that deviated noticeably from the majority have been removed from the data set. Depending on visual pre-selection, and further detection of unsuitable samples (called outliers) that were later found during modelling, different numbers of spectra for each parameter where used for the models. Details will be given later at each model description.

4.6.1 Pulp hand sheets

Model for M_w

The pre-processed (first derivative + multiplicative scattering correction) NIR-spectra in the wave number range from 7500 cm^{-1} to 4250 cm^{-1} and the molecular weight (M_w) determined for 98 samples by GPC (FDAM) were used to calculate a PLS-R model (73 spectra for calibration and cross validation) and to validate it (19 samples for test set validation). During modelling six samples (outliers) were removed from the calibration (cross validation) data set. The cross validation and test set validation results using 10 PLS components for the model are shown in figure 87, and the errors are listed in table 37.

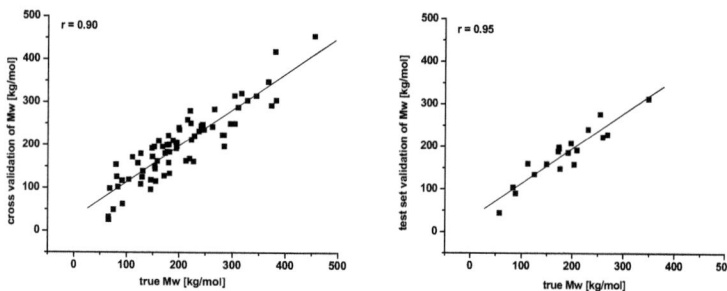

Figure 87. Cross validation (left) and test set validation (right) results of the molecular weight (M_w) of pulp hand sheets.

Model for carbonyl group content

From the 93 samples, with the carbonyl group content determined by fluorescence labelling using CCOA-label, 75 samples were used for calibration. The pre-processed (first derivative + multiplicative scattering correction, selected wave number ranges: 7155 – 5970 cm^{-1} + 5340 – 4945 cm^{-1} + 4500 - 3775 cm^{-1}) NIR-spectra were modelled by 7 PLS components. During cross validation the software detected six samples as outliers. They were removed. The test set validation with 18 samples used 7 PLS components to produce a prediction model (Figure 88). For errors see Table 36.

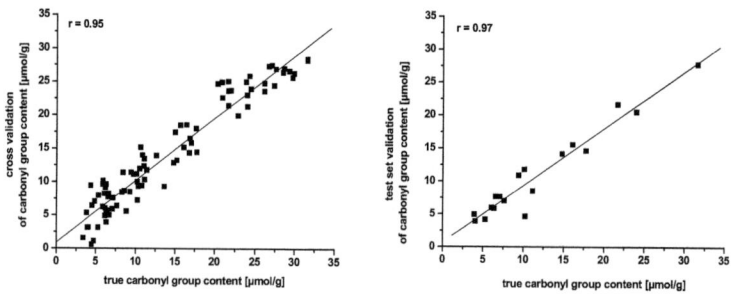

Figure 88. Cross validation (left) and test set validation (right) results of carbonyl group content of pulp hand sheets.

Model for carboxyl group content

Carboxyl group content was determined on 102 samples by fluorescence labelling using FDAM-label. The pre-processed (first derivative + vector normalization) NIR-spectra in the wave number range from 7525 - 3820 cm^{-1} and the carboxyl group content determined were used for modelling during which six samples (outliers) were removed. The cross validation containing 76 spectra and validated by test set validation (20 samples) using 10 PLS components are shown in figure 89. Errors are listed in table 36.

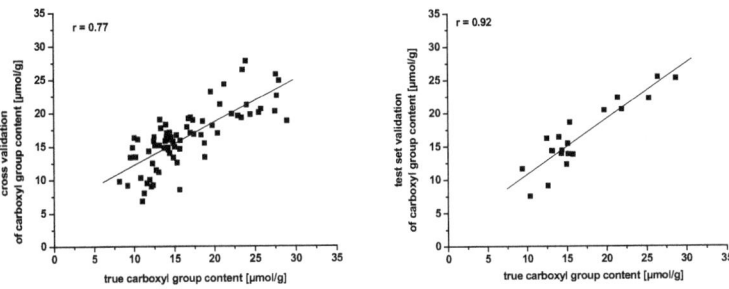

Figure 89. Cross validation (left) and test set validation (right) results of carboxyl group content of pulp hand sheets.

In Table 37, the results for the prediction of the M_w, carbonyl group content and carboxyl group content of pulp hand sheets are given. It is remarkable that test set validation yielded better results than cross validation. Especially for carboxyl group determination the cross validation model is not very satisfying at the first glance. Nevertheless, according to [183] it is possible to have a relatively bad cross validation while still a good prediction can be obtained. In this case it was decided to use the suggested model, not only because the satisfactory test set validation, but mainly because it was possible to use these parameters for rag papers as well (see section 4.6.2). Another advantage of these parameters was that the amount of samples that have to be removed is very low compared to other models tested.

Table 37. Parameters of cross validation (CV) and test set validation (TS) on pulp hand sheets

Parameter		Offset	Slope	r[1]	RMSECV[2] / RMSEP[3]
Molecular weight	CV	29 kg/mol	0.84	0.90	37 kg/mol
	TS	20 kg/mol	0.87	0.95	23 kg/mol
Carbonyl group content	CV	1.3 µmol/g	0.91	0.95	2.7 µmol/g
	TS	0.7 µmol/g	0.87	0.97	2.2 µmol/g
Carboxyl group content	CV	5.8 µmol/g	0.64	0.77	3.3 µmol/g
	TS	1.3 µmol/g	0.88	0.92	2.2 µmol/g

[1] r: correlation coefficient
[2] RMSECV: Root Mean Square Error of Cross Validation
[3] RMSEP: Root Mean Square Error of Prediction

4.6.2 Historic rag papers

Variations in colour, paper surface structure, paper thickness and visual appearance found within and among the analyzed paper were obviously quite big. While pulp hand sheets contain cellulose in a reasonable pure and uniform way, historic rag paper contain different fillers and sizing material, while also aging does its part in changing paper parameters and inflicting inhomogeneities in an uncontrolled way. Visual inspection of the rag paper spectra revealed considerably differences regarding their absorption even though several spectra represented one single reference value. In order to account for this, several spectra with almost identical reference values were averaged to give one spectra representing the desired parameter. Due to this averaging process of historic paper objects only a small number of samples (48 for M_w, 38 for carbonyl group content, 30 for carboxyl group content, and 31 for surface pH) could be used for the study. Therefore, no test set was available and the validation was restricted to cross validation of the models.

Figure 90 show the cross validation results of rag papers for the M_w, surface pH, carbonyl group content and carboxyl group content, respectively. Two samples for carboxyl group content, three samples for surface pH, and four samples for both, M_w and carbonyl group content were marked as outliers and removed during modelling.

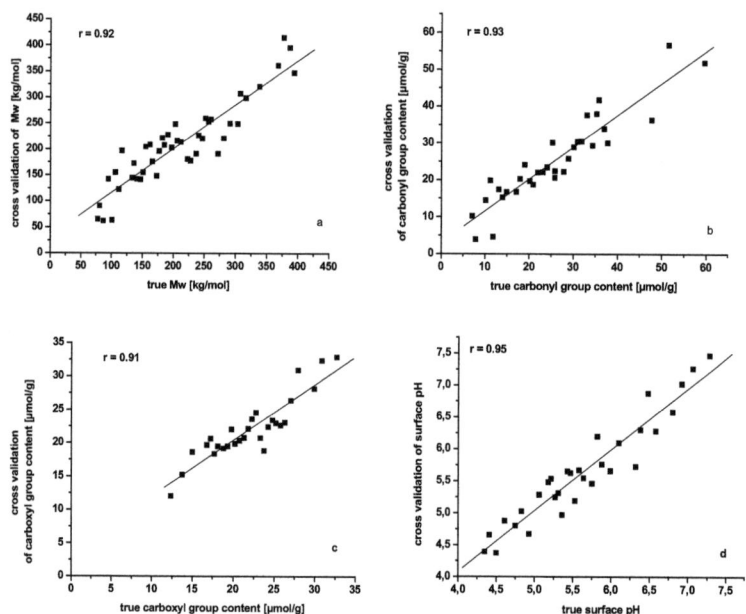

Figure 90. Cross validation results of molecular weight using 44 averaged spectra (a), carbonyl group content using 34 averaged spectra (b), carboxyl group content using 28 averaged spectra (c) and surface pH using 31 average spectra (d) of rag papers.

Despite the higher variability of these materials as described above the quality of the prediction was quite satisfying, being only slightly lower than for the pulp hand sheets. The averaging process that helps to level out inhomogeneities within the paper matrix explains this. The corresponding data are summarized in Table 38. Modelling of paper thickness and brightness did not lead to convincing models. No further optimization was attempted, because these parameters are already obtained in a non-destructive way.

Table 38. Parameters of cross validation on rag papers.

Parameter	Pre-processing	Wave number range [cm^{-1}]	Offset	Slope	r^1	RMSECV^2
M$_w$	Straight Line Subtraction	7500 - 6800	32 kg/mol	0.85	0.92	35 kg/mol
Carbonyl group content	First Derivative + Multiplicative Scattering Correction	7155 – 5970 + 5340 – 4945 + 4500 - 3775	3.0 μmol/g	0.87	0.93	4.7 μmol/g
Carboxyl group content	First Derivative + Vector Normalization	7525 - 3820	3.6 μmol/g	0.84	0.91	2.1 μmol/g
Surface pH	Multiplicative Scattering Correction	3845 – 5360 + 5850 - 7525	0.3	0.95	0.95	0.3

^{1} r: correlation coefficient
^{2} RMSECV: Root Mean Square Error of Cross Validation

For both oxidized functionalities, carbonyl and carboxyl group content, the same parameters were used for model building. Especially for carbonyl group content the obtained correlation is promising. Nevertheless, as for rag papers no test set validation was possible due to the strongly decreased sample amount because of spectra averaging, an attempt was made to combine pulp hand sheet and rag spectra for modelling. The results for cross validation and test set validation are shown in figure 91. Pre-processing (multiplicative scattering correction) and wave number selection (3775 – 7480 cm^{-1}) for the combined sample sets differ from those applied for the two types of sample sets when using them separately. Additionally, more PLS components were used (14) to model the data. The parameter obtained for the test set validation are comparable to those achieved in the cross validation of averaged rag paper samples and only slightly worse than those for pulp hand sheets (offset TS: 2.9 μmol/g, slope TS: 0.87, r TS: 0.90 and RMSEP: 5.0 μmol/g).

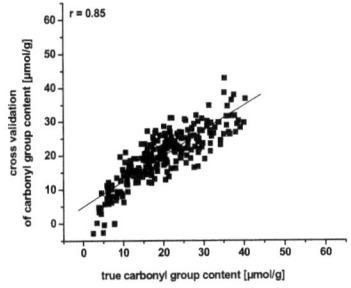 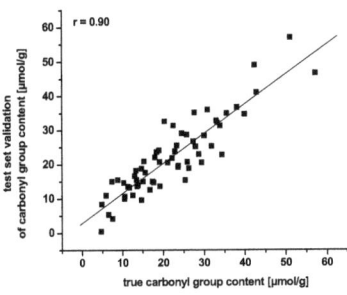

Figure 91. Cross validation (left) and test set validation (right) results of carbonyl group content of pulp hand sheets and rag papers without averaging.

4.6.3 Summary

NIR-spectra have the big advantage of being a non-destructive and fast means for pulp and paper evaluation. This fact is greatly appreciated in conservation dealing with unique artifacts that cannot be investigated the conventional, *i.e.* destructive way. When combined with multivariate calibration, quantitative evaluation of NIR-spectra becomes feasible. From literature it is known that especially lignin content expressed by the Kappa number [91, 184] and geographic origin of pulp as well as type of wood species used are reported to be predictable by NIR [185, 186], while so far no attempts have been made to correlate chemical properties such as carbonyl content and carboxyl content to NIR spectra.

In the present work, a simple, fast and non-destructive method based on NIR spectroscopy and multivariate data analysis for the determination of the molecular weight, carbonyl group content and carboxyl group content of pulp hand sheets and rag paper was developed for the first time. On rag papers also surface pH was made available. The quality of the obtained predictions was based on accurate reference methods (CCOA and FDAM).

The method allows testing on different sites of a paper to obtain the analytical parameters and gain insight into inhomogeneities of those parameters without damaging the paper. This is important for historic rag papers which have an inherently higher variability of the paper material, such as discolorations, varying thickness and surface textures. This deliberately incorporated variability could be modelled using average spectra of different historic paper samples as confirmed by cross validation of the key chemical parameters. Future work will be directed towards applying the model to unknown historic samples.

However, for every day life a simpler test is needed that does not depend on complicated models and expensive technical equipment that needs expertise and maintenance. Without any doubt, fluorescence labelling is a very appropriate technique to be the basis for a more straight forward correlation between precise analytical data and easy at hand condition rating in a workshop. Further research will focus on a method using a few fibres (therefore still being micro-destructive) and their behaviour towards easily available commercial chemicals to evaluate their actual condition. The role model for these tests should be already available micro-tests like the non-bleeding iron-test and similar ones, especially designed for the use in conservation workshops.

The most important drawback is possibly that multivariate calibration approach using infrared spectroscopy is technically challenging. Additionally the theoretical background of the calibration is very exigent and not easily understood. These facts are mentioned in many textbooks, and have lead experts to fight against the perception of a "black box" technique. Despite being based on spectroscopic rules and mathematical algorithms, it might rather be perceived as a confusing model facilitating doubtful applications and wrong conclusions. A lot of experience and experiments are needed to gain trust in the predictive ability of models.

5 Practical aspects

Even though the idea of scientific research is well accepted in the conservation community, it is often felt by manually working conservators that research work only deals with models that are not applicable for real life situations. The practical aspects helping to solve present problems in the conservation workshop often seem to be neglected. One might want to ask why such extensive research work using complicated analytics is necessary, and how it relates to practical work.

In addition to that, paper conservation has for a long time suffered from its dependence on mechanical tests developed for freshly prepared papers to assess their suitability for printing or packing purposes. How to estimate the condition using a test that makes historic papers fail within seconds at the lower limit of detection because it was designed for brand new papers? These test methods have been far too insensitive for paper conservation and additionally too material demanding. The testing of naturally aged and degraded papers was often out of scope. Even though these mechanical properties are linked to the key cellulose parameters molecular weight and molecular weight distribution, the evaluation of oxidized functionalities, namely carbonyl and carboxyl groups, is out of reach.

Thus, by means of mechanical testing, viscosity and pH measurements, probably the three most widely spread and accepted analysis in paper conservation, the direct elucidation of degradation pathways and the evaluation of treatment effectiveness is limited. Firstly, the parameters acquired are mostly unspecific. Mechanical weakness and a low DP are the product of degradation, but not its cause. A low pH certainly contributes to decreased stability as it promotes acid hydrolysis, but the influence of oxidation is not detected. Secondly, standard procedures of the three methods described above require one or more grams of sample material. This impedes a spatial resolution of damage and the analysis of originals to compare simulated damage with naturally occurring.

These general statements on available and relatively well known test methods and their shortcomings underline the importance of more sophisticated chemical analysis. Still, research work has to be liked to the conservation workshop, a task that is even more difficult with more complex instrumental set-ups. About the relationship between conservation and science, the following was discussed [187]:

- Conservation should attempt to preserve or restore the true nature of objects, while science helps to assess a given object's condition, its history and a chosen treatment on its suitability. This is important because in a western context an object's true nature relies mainly upon its material constituents – therefore replications are not considered to be comparable to originals – and therefore much knowledge about the material itself has to be included in conservation work.

- The techniques to be used and the desired target state of the conservation process should be determined by scientific means. These scientific means can be defined by terms like quality control (What was the aim of the work and in how far was it achieved?) and objectivity (What are the arguments - based on experience, knowledge and analytical data - that lead to a certain conclusion and decision?). Decisions should be based upon objective facts and explained according to them. The idea of artistic inspiration is not accepted in this concept.

- All scientific data and results need interpretation. They have to be embedded into a given context that is out of scope of analytical methods. Relevant and important classifications in conservation like "damage" that needs to be stopped and "patina" that should be preserved are based on personal judgement of experienced people. The underlying criteria of what to classify as damage or patina cannot be determined in an analytical way. Therefore cooperation between several experts is needed, even though time consuming and difficult to achieve.

To conclude, mechanisms of degradation have to be explained by scientists to gain more confidence about how to act towards them. The quantification and precise description of the extent of damage, given in numbers, will lead to the choice of an appropriate treatment based on well justified understanding. To define how this treatment should be applied to which object is the task of the conservator. Modifications and limitations of the methods have to be topic of discussion, based on objective arguments.

This might be demonstrated by the relatively wide acceptance of deacidification treatments. Even though far from being completely understood in all its consequences a general knowledge on the prevailing degradation mechanism of paper was acquired (acid hydrolysis), a method that counteracted this mechanism was developed based on observations and experiments (deacidification), elaborated (mass deacidification technology and further improvements for aqueous deacidification in small scale) and today this treatment concept and its necessity is accepted and spread far beyond the boundaries of the conservation community. The chapter about the sustainability of mass deacidification in this work contributes to the general understanding of the interplay between beneficial and possibly detrimental effects of this treatment.

The same contribution for further understanding of phenomena like cellulose degradation in general and accelerated degradation caused by transition metal ions, was pursuit in the present work. Being able to quantify the degree and the underlying mechanism and to partly simulate them should encourage treatments that directly address the causes of damage and efficiently counteract them. Using GPC-multi detector set-up gives scientists working in paper conservation a tool at hand that can be used to study paper conservation related topics with small amounts of sample material, sensitive and precise results that are additionally directly linked to molecular key properties of cellulose and partially hemicelluloses, too.

A very illustrative example of the desired cooperation between research laboratories and conservation workshops is the evaluation of the calcium phytate/ calcium hydrogen carbonate treatment to inhibit degradation of cellulose caused by irongall ink. Based on earlier knowledge, a method was chosen and adapted to the need of a specific collection selected for treatment by a team of conservators. This expertise assures that the treatment is indeed feasible for practical conservation work. Accompanying analysis, partly performed in the framework of this study, helped to prove the effectiveness of the chosen treatment on a wide range of materials. Additionally, in the present work the underlying degradation mechanism was revealed and classified on historic papers, and the synergistic effect of irongall ink was proven, unambiguously underlining the importance of a combined treatment.

6 Summary

General

The main goal of the present study was to describe and evaluate paper conservation topics by fluorescence labelling of cellulose functionalities prior to GPC for the determination of molecular weight and molecular weight distribution and profiles of functional groups. Further emphasis was put on improving the solubility of some pulps and on complementing the above described method by a non-destructive approach.

The method of fluorescence labelling followed by GPC was developed for the analysis of dissolving pulps in the production of fibres from renewable sources. It was found that the method could be adopted for rag papers and a relatively broad variety of modern, 19th and 20th century papers despite of the presence of sizing and filling material. Even small amounts of lignin as present in 19th and 20th century material are tolerated by the method.

Of special interest for paper conservation is the fact, that very small amounts of sample material are sufficient. Down scaling to about 2 mg is possible, 5 - 25 mg of sample run on a routine basis. Even though still being destructive – the sample material is dissolved, fluorescence labelled, measured and thereafter discarded – the low sample amount allows for analysis of selected original material when sampling on unsuspicious areas is possible or loose fragments are available. Additionally, the spatial distribution of damage in the paper under investigation can be visualized when multi sub sampling is allowed

Fluorescence labelling followed by GPC is an excellent method for the analysis of many conservation relevant topics as shown in this study. Several topics have been studied among them general aging characteristics, the detrimental action of transition metal ions on paper and treatment possibilities against it, the influence of wet dry interfaces on cellulose and the sustainability of mass deacidification treatments. All topics that need to detect cellulose degradation in terms of changes on the molecular weight of cellulose and on cellulose functionalities can be studied in detail by fluorescence labelling and subsequent GPC.

Non-destructive approach

After assuring the appropriateness of the fluorescence labelling method for historic papers as found in paper conservation by measuring a broad variety of pulps and historic papers from different sources and in different conditions, this method was chosen as basis for the development of a non-destructive approach of paper analysis by multivariate calibration and near infrared spectroscopy. Different models for pulp hand sheets and historic rag papers have been calibrated and validated for each of the desired parameters, molecular weight, carbonyl and carboxyl group content. For rag papers also surface pH was made available. It is possible to model naturally occurring variances of pulp and historic rag papers into a model that can predict the desired properties just by taking an infrared spectrum. No model for ground wood papers was achieved, basically because this type of paper is not accessible by the main dissolving agent used before GPC-analysis, DMAc/ LiCl 9%. Therefore no reliable data have been obtained that are needed for model calibration.

Improving cellulose solubility

A broad variety of 19th and 20th century papers are not soluble in DMAc/ LiCl, mainly due to their high lignin content. Consequently their molecular weight and oxidized functionalities cannot be determined via fluorescence labelling and GPC. A well specified example of such an insoluble paper was chosen as a standard paper to test various approaches to improve solubility in DMAc/ LiCl or similar systems. The most promising suggestion was the combination of direct solution by DMAc/ LiCl and derivatization by EIC. Even though the quantification of oxidized functionalities was still possible after combining the two methods, the selected standard paper could not be dissolved. Some success was achieved when replacing DMAc by DMI, and when extraction with unpolar solvents was performed prior to addition of DMAc/ LiCl. However no complete solution was achieved during this study.

Aging characteristics

The broad analysis of papers used for the set-up of the above mentioned non-destructive method made general knowledge available about the condition of naturally aged rag papers. Most of them are made out of linen and hemp fibres till the intensified use of cotton at the beginning of the 19th century [188] and the general substitution of recycled textiles and annual plants by ground wood pulp and later chemical pulps starting in the middle of the 19th century. Considering the relatively narrow distribution of values to be expected in this type of papers as shown in the present study, the parameters for typical European rag paper after natural aging have been defined. Within a certain range of probabilities other naturally aged rag papers are expected to have the same parameters.

Influence of transition metal ions

The topic that was studied most during this study was degradation of cellulose caused by transition metal ions, mainly by iron and copper ions or combinations thereof. For both metal ions it was found that damage is not limited to the ink or pigment line. In contrast, it can be traced within significant distances in mm range. This is especially true for copper ions. Due to the fact, that copper acetate pigments are far more soluble than the irongall ink complex, most severe damage was even found next to the pigment application in some cases. Iron gall ink always caused most damage directly below the ink line. It was demonstrated that copper pigments will increase the overall amount of copper ions in a sample far more than iron ions. This implies that paper suffering from the influence of copper ions will benefit more from aqueous treatments as the overall copper ion content will be reduced throughout the paper. However, this suggestion contrasts to the results from aqueous treatment of simulated copper corrosion when especially the aqueous treatments lead to worst degradation.

The application of ink or pigment can usually be traced in the molecular weight distribution because the lower molecular weight fractions gain more importance when cellulose suffers from the influence of transition metal ions. Copper ions lead to a more severe change of the molecular weight distribution, turning it from a generally found bimodal MWD into a trimodal one, a fact that has not been observed in irongall ink corrosion so far.

It was also possible for the first time to describe the underlying degradation mechanism of historic papers suffering from the influence of iron gall ink or copper pigments. Even though it has always been postulated that iron gall ink corrosion and copper corrosion are caused by oxidative processes it has scarcely been proven on historic samples material. Being able to analyse very small sample amounts the spatial resolution of degradation and direct assignment to a certain degradation process became available. As expected, in most cases the damage was traced back to an oxidative action with increased oxidized cellulose functionalities and increased chain scission like it has been postulated. Some exceptions of that general rule were found when samples were mainly degraded by hydrolytic action and additional oxidized functionalities have been introduced mostly due to newly formed reducing end groups after chain scission. The fact, that sometimes the degradation is rather based on acid hydrolysis explains the partial success of deacidification treatments on papers damaged by irongall ink.

Some of the above described observations are reproducible on simulated models after accelerated aging. For copper corrosion oxidative damage was successfully achieved on model rag papers that altered cellulose most next to the pigment application. The formation of a low molecular weight shoulder in these samples was also observed. When iron gall ink was applied on Whatman filter paper, increased brittleness was achieved and also a low molecular weight shoulder. On the other hand it was not possible to induce strong differences between water absorbing and hydrophobic areas, neither on copper pigment nor on irongall ink containing models. On model rag papers the damage caused by iron gall inks only proceeded very slowly. This outcome has lead to the conclusion that the simulation of irongall corrosion can still be improved. The simulation of certain degradation patterns is not only desirable in order to study underlying mechanisms, but also when designing a treatment procedure. Some typical characteristics of historical papers damaged by irongall ink have been defined:

- Increase of the low molecular weight shoulder
- Different behaviour of pure paper and ink/ pigment line towards water
- Mechanical weakness that is directly linked to low average molecular weight

These parameters should be achieved in model papers after accelerated aging.

Another important application for fluorescence labelling in paper conservation studies is the possibility to test the impact of conservation treatments. In the case of transition metal ions the effectiveness of calcium phytate/ calcium hydrogen carbonate treatment on irongall inks and irongall inks containing additional copper sulphate was successfully proven on model papers and historic samples. As fluorescence labelling detects all key parameters of cellulose it was possible to show, that oxidation was suppressed and chain scission stopped. All data were obtained on sufficient significant level. This is a main advantage towards standard mechanical tests that are usually not sensitive enough to detect small differences and additionally have big standard deviations that impede statistically significant differentiation between different treatment options tested. As the addition of copper ions to the model ink and the presence of copper ions in historic samples did not lower the effectiveness of calcium phytate/ calcium hydrogen carbonate treatment, further research should be dedicated to applying this treatment option on copper corrosion as well.

Deacidification

Another conservation treatment tested was the deacidification of 19th and 20th century papers. Even though not all of these papers, namely those having high ground wood content, were available for analysis, appropriate sample material was found to gain some general insight into the behaviour of this type of papers and their performance towards deacidification treatment. It seems that deacidification treatment provides papers with more homogenous molecular weight parameters. While non-treated papers appear to be very inhomogeneous, treated papers have more similar values of M_w. Even though no proof was found that inhomogeneities in pH will cause inhomogeneous aging, a more unified distribution is still considered to be a desirable treatment result. Mapping of several papers showed that historic papers will feature different zones of degradation, making homogeneous results more difficult to achieve.

Next to the question of successful deacidification by neutralization of acids and deposition of an alkaline reserve the question was raised in how far β-elimination could play a role in the application of a highly alkaline substance on oxidized papers. Especially the role of magnesium ions is still unsolved in mass deacidification. Magnesium compounds are often criticized for their strongly alkaline pH and their tendency to cause colour changes in treated papers. On pulp hand sheets with sufficiently high carbonyl group content β-elimination was proven. However, even in pure cellulose samples a certain amount of carbonyl groups had to be exceeded before β-elimination gained any importance at all. The detrimental influence of β-elimination on 20th century papers with fillers and sizing agents was difficult to display. Generally the prevailing beneficial aspect of deacidification was demonstrated.

Wet-dry interfaces

Magnesium ions were found to be of some importance in a completely different field of research. When studying the condition of cellulose in wet dry interfaces neither increased chain scission nor increased carbonyl or carboxyl group content was found in Whatman filter papers with artificially produced tidelines even after several days of accelerated aging. Contrary to that, especially the zones of filter paper immersed into water for several hours in order to produce the interfaces suffered from measurable amounts of damage. In some cases it was even found that the cellulose molecules of the interface itself were in remarkably good condition, even better than the zone of filter paper that was not influenced by water immersion and measured as a reference for unchanged cellulose. Further analysis revealed that magnesium ions were accumulated in the wet dry interface of Whatman filter paper models. Other samples that were produced under normal lab conditions without special care for cleanness exhibited large amounts of copper ions accumulated in the tideline region. This might explain increased amounts of peroxides found in the tideline region. When analysing historic wet dry interfaces none of these accumulations of metal ions were found. Metal ions do not seem to play any important role in the formation of these brown lines.

Whatman filter paper and historic papers have in common that fluorescent material is generated or conducted by water to the tideline region. This typical fluorescence is found to be reduced in the area that was influenced by water. Also in several naturally aged papers with tidelines on them no general trend for significantly more degradation or oxidation was found in the tideline zone. However, it should be emphasized that short chained cellulose degradation products are not detected by GPC-MALLS. Their possible presence though did not exert any detectable detrimental influence on the cellulose molecule. This has lead to the conclusion that wet dry interfaces might rather be an esthetical than a conservation problem and that measurements taken for their removal will more seriously harm the cellulose than simply leaving the marks.

7 References

1. Fengel, D.; Wegener, G., *Wood: Chemistry, Ultrastructure, Reactions*. de Gruyter: Berlin, 1989.
2. Klemm, D.; Schmauder, H. P.; Heinze, T., Cellulose. In *Polysaccharides II - Polysaccharides from Eukaryotes*, De Baets, S., Vandamme, E.J., Steinbüchel, A., Ed. Wiley-VCH: Weinheim, 2002; Vol. 6, pp 275-319.
3. Beazley, K., Mineral Fillers in Paper. *The Paper Conservator* **1991**, 15, 17-27.
4. Klemm, D.; Philipp, B.; Heinze, T.; Heinze, U.; Wagenknecht, W., *Comprehensive Cellulose Chemistry. Volume 1: Fundamentals and Analytical Methods*. 1998.
5. Kroon-Batenburg, L. M. J.; Kroon, J.; Nordholt, M. G., Chain Modulus and Intramolecular Hydrogen Bonding in Native and Regenerated Cellulose Fibres. *Polymer Communications Guiltford* **1986**, 27, 290-292.
6. Ebringerová, A.; Hromádková, Z.; Heinze, T., Hemicellulose. *Advances in Polymer Science* **2005**, 186, 1-67.
7. Ebringerova, A., Structural Diversity and Application Potential of Hemicelluloses. *Macromolecular Symposia* **2006**, 232, 1–12.
8. Koch, G., Raw Material for Pulp. In *Handbook of Pulp*, Sixta, H., Ed. Wiley-VCH: Weinheim, 2006; Vol. 1, pp 21-68.
9. Correia, F.; Roy, D. N., Analysis of Hemp Chemical Pulp Monosaccharide Degradation Compared with Aspen and Spruce Chemical Pulps. *Journal of Natural Fibers* **2005**, 2, 35-58.
10. Spiegelberg, H. L., The effect of hemicelluloses on the mechanical properties of individual pulp fibres. In *Tappi Journal*, 1966; Vol. 49, pp 388-396.
11. Luner, P., Paper Permanence. *Tappi Journal* **1969**, 52, 796-805.
12. Potthast, A., Chemistry of (Acid) Sulfite Pulping. In *Handbook of Pulp*, Sixta, H., Ed. Wiley-VCH: Weinheim, 2006; Vol. 1, pp 405-427.
13. Sakakibara, A.; Sano, Y., Chemistry of Lignin. In *Wood and Cellulosic Chemistry*, 2 ed.; Hon, D. N.-S., Shiraishi, N., Ed. Marcel Dekker: New York, 2001; pp 109-173.
14. Zou, X.; Gurnagul, N.; Deschatelets, S.; Begin, P.; Iraci, J.; Grattan, D.; Kaminska, E.; Woods, D. In *Proceedings of the Canadian co-operative permanent paper research project: The impact of lignin on paper permanence*, Process and Product Quality Conference and Trade Fair 1998; TAPPI Press, Norcross, GA, United States: 1998; pp 65-74.
15. Potthast, A.; Schiehser, S.; Rosenau, T.; Sixta, H.; Kosma, P., Effect on UV radiation on the carbonyl distribution in different pulps. *Holzforschung* **2004**, 58, 597-602.
16. Daniels, V. D., The Chemistry of Paper Conservation. *Chemical Society Reviews* **1996**, 179-186.
17. Porck, H. J. In *The Bookkeeper Process and its application at the National Library of the Netherlands*, Save Paper! Mass deacidification. Today's Experiences - Tomorrow's Perspectives, Bern, 2006; Blüher, A., Grossenbacher, G., Banik, G., Ed. Swiss National Library: Bern, 2006; pp 37-42.
18. Barrow, W. J.; Sproull, R. C., Permanence in Book Papers. *Science* **1959**, 129, (3356), 1075-1084.
19. Sixta, H., Sulfite Chemical Pulping. In *Handbook of Pulp*, Sixta, H., Ed. Wiley-VCH: Weinheim, 2006; Vol. 1, pp 392-405.
20. Lewin, M.; Mark, H. F., Oxidation and Ageing of Cellulose. *Macromolecular Symposia* **1997**, 118, 715-724.
21. Zou, X.; Uesaka, T.; Gurnagul, N., Prediction of Paper Permanence by Accelerated Aging. Part 1: Kinetic Analysis of the Aging Process. *Cellulose* **1996**, 3, 243-267.
22. Feller, R. L., Accelerated Aging. Photochemical and Thermal Aspects. **1994**.

23. Dessauer, G., Die Probleme und Ursachen des Alterns beim Papier und die neuesten Möglichkeiten der Papierindustrie, alterungsbeständige Papiere zu erzeugen. *Das Papier* **1980**, 3, 249-255.
24. Davidson, R. S., The photodegradation of some naturally occurring polymers. *Journal of Photochemistry and Photobiology B: Biology* **1996**, 33, 3-25.
25. Maier, C.; Petersen, K., *Schimmelpilze auf Papier. Ein Handbuch für Restauratoren. Biologische Grundlagen, Erkennung, Behandlung und Prävention*. Der andere Verlag: Tönning, 2006.
26. Havermans, J. B. G. A., Effects of Air Pollutants on the accelerated Ageing of cellulose-based Materials. *Restaurator* **1995**, 16, 209-233.
27. Daniel, F.; Flieder, F.; Leclerc, F., Study of the Effects of Pollution on Deacidified Papers. *Les Documents Graphique et Photographiques: Analyse et conservation, Travaux du Centre de Recherches sur la Conservation des Documents Graphique* **1988**, 53-92.
28. Padfield, T., Conservation Physics. In *Stress, strain and craquelure*, http://www.padfield.org/tim/cfys/.
29. Bogaard, J.; Whitmore, P. M. In *Explorations of the Role of Humidity Fluctuations in the Deterioration of Paper*, Works of Art on Paper. Books, Documents and Photographs. Techniques and Conservation, Baltimore, 2002; Daniels, V. D., Dornithorne, A., Smith, P., Ed. International Institute for Conservation of Historic and Artistic Works: Baltimore, 2002; pp 11-14.
30. Porck, H. J. *Rate of Paper Degradation: The Predictive Value of Artificial Aging Tests*; European Commission on Preservation and Access: Amsterdam, 2000; pp 1-40.
31. Wilson, W. K.; Harvey, J. L.; Mandel, J.; Worksman, T., Accelerated Aging of Record Papers Compared with Normal Aging. *Tappi Journal* **1955**, 38, 543-548.
32. Letnar, M. C.; Vodopivec, J., Influence of paper raw materials and technological conditions of paper manufacture on paper ageing. *Restaurator* **1997**, 18, 73-91.
33. Neevel, J. G., The ink corrosion project at NICH a review. *Abbey Newsletter* **1997**, 21, 88-92.
34. Barrett, T. D., Early European Papers/Contemporary Conservation Papers. A Report on Research Undertaken from Fall 1984 through Fall 1987. *The Paper Conservator* **1989**, 13, 57-65.
35. Inaba, M.; Sugisita, R., The Deterioration of Japanese Paper by Accelerated Aging Treatment. In *Third International Institute of Paper Conservation Conference*, University of Manchester Institute of Science and Technology, 1992.
36. Banik, G.; Sobotka, W. K.; Vendl, A.; Norzsicska, S., Effects of Atmospheric Pollutants on Deacidified Modern Papers. In *10th Triennial Meeting*, Bridgland, J., Ed. ICOM Committee for Conservation, Paris: Washington DC, USA, 1993; pp 435-441.
37. Pedersoli, J. L. In *The development of micro-analytical methodologies for the characterization of the condition of paper,*, 9th International Congress of IADA, Copenhagen, August 15th – 21st 1999, 1999; Copenhagen, 1999; pp 107-114.
38. Porck, H. J.; Teygeler, R. *Preservation Science Survey. An Overview of Recent Developments in Research on the Conservation of Selected Analog Library and Archival Materials*; European Commission on Preservation and Access: Amsterdam, 2001; pp 1-79.
39. Parisot, A.; Cyrot, J., The Degree of Polymerization of Cellulose. *Journal of the Textile Institute* **1951**, 42, 783-797.
40. Cael, J. J.; Cietek, D. J.; Kolpak, F. J., Application of GPC/ LALLS to Cellulose Research. *Applied Polymer Symposia* **1983**, 509-529.
41. Segal, L., Characterization of Celluloses by Gel Permeation Chromatography. *Journal of Polymer Sciences: Part C* **1986**, 21, 267-282.

42. Dupont, A.-L.; Mortha, G., Comparative Evaluation of Size-Exclusion Chromatography and Viscosimetry for the Characterisation of Cellulose. *Journal of Chromatography A* **2004**, 1026, 129-141.
43. Evans, R.; Wearne, R. H.; Wallis, A. F. A., Effect of amines on the carbanilation of cellulose with phenylisocyanate. *Journal of Applied Polymer Science* **1991a**, 42, 813-820.
44. Evans, R.; Wearne, R. H.; Wallis, A. F. A., Pyridine-catalyzed depolymerization of cellulose during carbanilation with phenylisocyanate in dimethylsulfoxide. *Journal of Applied Polymer Science* **1991b**, 42, 821-827.
45. Fischer, M.; Fischer, K., Polymer-analogous preparation of cellulose tricarbanilates: Mechanisms of degradation in dimethylsulfoxide. *Macromolecular Symposia* **2005**, 223, 121-135.
46. McCormick, C. L.; Callais, P. A., Derivatization of cellulose in lithium chloride and N,N-dimethylacetamide solutions. *Polymer* **1987**, 28, 2317-2323.
47. Dupont, A.-L., Cellulose in Lithium Chloride/ N,N-dimethylacetamide, Optimisation of Dissolution Method Using Paper Substrates and Stability of the Solutions. *Polymer* **2003**, 44, 4117-4126.
48. Potthast, A.; Rosenau, T.; Sixta, H.; Kosma, P., Degradation of cellulosic materials by heating in DMAc/LiCl. *Tetrahedron Letters* **2002**, 43, 7757-7759.
49. Berggren, R.; Berthold, F.; Sjöholm, E.; Lindström, M., Improved methods for evaluating the molar mass distributions of cellulose in kraft pulp. *Journal of Applied Polymer Science* **2003**, 88, 1170-1179.
50. Potthast, A.; Rosenau, T.; Buchner, R.; Röder, T.; Ebner, G.; Bruglachner, H.; Sixta, H.; Kosma, P., The cellulose solvent system N,N-dimethylacetamide/lithium chloride revisited: The effect of water on physicochemical properties and chemical stability. *Cellulose* **2002**, 9, 41-53.
51. Sjöholm, E.; Gustafsson, K.; Pettersson, B.; Colmsjö, A., Characterization of the cellulosic residues from lithium chloride/N,N-dimethylacetamide dissolution of softwood kraft pulp. *Carbohydrate Polymers* **1997**, 32, 57-63.
52. Berthold, F.; Gustafsson, K.; Berggren, R.; Sjöholm, E.; Lindström, M., Dissolution of softwood kraft pulps by direct derivatization in lithium chloride/N,N-dimethylacetamide. *Journal of Applied Polymer Science* **2004**, 94, 424-431.
53. Arndt, K. F.; Müller, G., *Polymercharakterisierung*. Hanser: München, 1996.
54. Mori, S.; Barth, H. G., *Size Exclusion Chromatography*. Springer: Heidelberg, 1999.
55. Wyatt, P. J., Light Scattering and the absolute characterization of macromolecules. *Analytica Chimica Acta* **1993**, 272, 1-40.
56. Potthast, A.; Rosenau, T.; Kosma, P., *Analysis of Oxidized Functionalities in Cellulose*. Springer: Heidelberg, 2006; Vol. 205, p 1-48.
57. Potthast, A.; Rosenau, T.; Kosma, P.; Saariaho, A.-M.; Vuorinen, T., On the Nature of Carbonyl Groups in Cellulosic Pulps. *Cellulose* **2005**, 12, 43-50.
58. Gruber, E., Analytical Characterisation of Pulps. In *Handbook of Pulp*, Sixta, H., Ed. Wiley-VCH: Weinheim, 2006; Vol. 2, pp 1211-1284.
59. Potthast, A.; Röhrling, A.; Rosenau, T.; Borgards, A.; Sixta, H.; Kosma, P., A novel method for the determination of carbonyl groups in cellulosics by fluorescence labeling. 3. Monitoring oxidative processes. *Biomacromolecules* **2003**, 4, 743-749.
60. Röhrling, J.; Potthast, A.; Rosenau, T.; Lange, T.; Borgards, A.; Sixta, H.; Kosma, P., A novel method for the determination of carbonyl groups in cellulosics by fluorescence labeling. 2. Validation and applications. *Biomacromolecules* **2002**, 3, 969-975.
61. Röhrling, J.; Potthast, A.; Rosenau, T.; Lange, T.; Ebner, G.; Sixta, H.; Kosma, P., A novel method for the determination of carbonyl groups in cellulosics by fluorescence labeling. 1. Method development. *Biomacromolecules* **2002**, 3, 959-968.

62. Fardim, P.; Holmbom, B.; Ivaska, A.; Karhu, J.; Mortha, G.; Laine, J., Critical comparison and validation of methods for determination of anionic groups in pulp fibres. *Nordic Pulp and Paper Research Journal* **2002**, 17, 346-351.
63. Bhardwaj, N. K.; Dang, V. Q.; Nguyen, K. L., Determination of carboxyl content in high-yield kraft pulps using photoacoustic rapid-scan Fourier transform infrared spectroscopy. *Analytical Chemistry* **2006**, 78, 6818-6825.
64. Bohrn, R.; Potthast, A.; Schiehser, S.; Rosenau, T.; Sixta, H.; Kosma, P., The FDAM method: Determination of carboxyl profiles in cellulosic materials by combining group-selective fluorescence labeling with GPC. *Biomacromolecules* **2006**, 7, 1743-1750.
65. Saverwyns, S.; Sizaire, V.; Wouters, J., The acidity of Paper. Evaluation of methods to measure the pH of paper samples. *13th Triennial Meeting, ICOM Committee for Conservation, Graphic Documents* **2002**, vol. II, 628-634.
66. Strlič, M.; Kolar, J.; Kočar, D.; Drnovšek, T.; Šelih, V. S.; Susič, R.; Pihlar, B., What is the pH of alkaline paper? *e-preservationScience* **2004**, 1, 35-47.
67. Stol, R.; Pedersoli, J. L.; Poppe, H.; Kok, W. T., Application of size exclusion electrochromatography to the microanalytical determination of the molecular mass distribution of celluloses from objects of cultural and historical value. *Analytical Chemistry* **2002**, 74, 2314-2320.
68. Rychlý, J.; Strlič, M.; Matisova-Rychlá, L.; Kolar, J., Chemiluminescence from paper I. Kinetic analysis of thermal oxidation of cellulose. *Polymer Degradation and Stability* **2002**, 78, 357-367.
69. Kelly, G. B.; Williams, J. C.; Mendenhall, G. D.; Ogle, C. A., Use of Chemiluminescence in the Study of Paper Permanence. *ACS Symposium Series* **1978**, (95), 117-125.
70. Strlič, M.; Kolar, J.; Pihlar, B.; Rychlý, J.; Matisova-Rychlá, L., Chemiluminescence during thermal and thermo-oxidative degradation of cellulose. *European Polymer Journal* **2000**, 36, 2351-2358.
71. Strlič, M.; Kočar, D.; Kolar, J., *Chemiluminometry of Cellulosic Materials*. Amercian Chemical Society: Washington, D.C., 2007; Vol. 954, p 531-545.
72. Strlič, M.; Kočar, D.; Kolar, J.; Rychlý, J.; Pihlar, B., Degradation of Pullulans of Narrow Molecular Weight Distributions - the Role of Aldehydes in the Oxidation of Polysaccharides. *Carbohydrate Polymers* **2003**, 54, 221-228.
73. Matisová-Rychlá, L.; Rychlý, J.; Slovák, K., Effect of the polymer type and experimiental parameters on chemiluminescence curves of selected materials. *Polymer Degradation and Stability* **2003**, 82, 173-180.
74. Doering, T. Altes Papier und Neue Techniken - Zerstörungsfreie Untersuchungen von Papier mit Festphasenextraktion (SPME). Technische Universitat Stuttgart, Stuttgart, 2006.
75. Doering, T.; Fischer, P.; Binder, U.; Liers, J.; Banik, G. In *An Approach to Evaluate the Condition of Paper by a Non-destructive Analytical Method*, Advances in Printing Science & Technology, Graz, 2000; Bristow, A., Ed. Leatherhead: Graz, 2000; pp 27-39.
76. Lattuati-Derieux, A.; Bonnassies-Termes, S.; Lavédrine, B., Identification of volatile organic compounds emitted by a naturally aged book using solid-phase microextraction/gas chromatography/mass spectrometry. *Journal of Chromatography A* **2004**, 1026, 9-18.
77. Hesse, M.; Meier, H.; Zeeh, B., Infrared and Raman Spectroscopy. In *Spectroscopic Methods in Organic Chemistry*, Georg Thieme Verlag: Stuttgart, 1997; pp 29-70.
78. Forsskahl, I.; Kentta, E.; Kyyronönen, P.; Sundström, O., Depth profiling of a photochemically yellowed paper: FTIR techniques. *Applied Spectroscopy* **1995**, 49, 163-170.
79. van Bronswijk, W.; Kirwan, L. J.; Fawell, P. D., In situ adsorption densities of polyacrylates on hematite nano-particle films as determined by ATR-FTIR spectroscopy *Vibrational Spectroscopy* **2006**, 41, 176-181.
80. Workman, J. J., Infrared and Raman Spectroscopy in pulp and paper analysis. *Applied Spectroscopy Reviews* **2001**, 36, 139-168.

81. Sistach, M. C.; Ferrer, N.; Romero, M. T., Fourier Transform Infrared Spectroscopy Applied to the Analysis of Ancient Manuscripts. *Restaurator* **1998,** 19, 173-186.

82. Calvini, P.; Gorassini, A., The Degrading Action of Iron and Copper on Paper: A FTIR-Deconvolution Analysis. *Restaurator* **2002,** 23, 205-221.

83. Dupont, A.-L., Degradation of cellulose at the wet/dry interface: II. An approach to the identification of the oxidation compounds. *Restaurator* **1996,** 17, 145-164.

84. Łojewska, J.; Miśkowiec, P.; Łojewski, T.; Proniewicz, L. M., Cellulose oxidative and hydrolytic degradation: In situ FTIR approach. *Polymer Degradation and Stability* **2005,** 88, 512-520.

85. Dufour, J.; Havermans, J. B. G. A., Study of the photo-oxidation of mass-deacidified papers. *Restaurator* **2001,** 22, 20-40.

86. Gibert Vives, J.; Daga Monmany, J.; Areal Guerra, R., Non-destructive method for alkaline reserve determination in paper. Diffuse Reflectance Infrared Fourier Transform Spectroscopy. *Restaurator* **2004,** 25, 47-67.

87. Hon, D. N.-S., Fourier Transform IR Spectroscopy and Electron Spectroscopy for Chemical Analysis: Use in the Study of Paper Documents. *Historic Textile and Paper Materials* **1986,** 212, 349-361.

88. Ali, M.; Emsley, A. M.; Herman, H.; Heywood, R. J., Spectroscopic studies of the ageing of cellulosic paper. *Polymer* **2001,** 42, 2893-2900.

89. Fardim, P.; Ferreira, M. M. C.; Duran, N., Multivariate calibration for quantitative analysis of eucalypt kraft pulp by NIR spectrometry. *Journal of Wood Chemistry and Technology* **2002,** 22, 67-81.

90. Trafela, T.; Strlic, M.; Kolar, J.; Lichtblau, D. A.; Anders, M.; Mencigar, D. P.; Pihlar, B., Nondestructive analysis and dating of historical paper based on IR Spectroscopy and chemometric data evaluation. *Analytical Chemistry* **2007,** 79, 6319-6323.

91. Alves, A.; Santos, A.; Da Silva Perez, D.; Rodrigues, J.; Pereira, H. a.; Simões, R.; Schwanninger, M., NIR PLSR model selection for Kappa number prediction of maritime pine Kraft pulps. *Wood Science and Technology* **2007,** 41, 491-499.

92. Siesler, H. W.; Ozaki, Y.; Kawata, S.; Heise, H. M., Near Infrared Spectroscopy - Principles, Instruments, Applications. *Wiley-VCH, Weinheim* **2002**.

93. Siesler, H. W. In *Vibrational spectroscopy of polymers. Analysis, physics and process-control*, Polymeric Materials Science and Engineering, , 1991; ACS, Washington, DC, United States: 1991; p 30.

94. Geladi, P., Kowalski, B.R., Partial Least Squares Regression: A Tutorial. *Analytica Chimica Acta* **1986,** 185, 1-17.

95. Beebe, K. R.; Pell, R. J.; Seasholtz, M. B., *Chemometrics: A practical guide*. Wiley Interscience: New York, 1998.

96. Brandis, L.; Lyall, J., Properties of Paper in Naturally Aged Books. *Restaurator* **1997,** 18, 115-130.

97. Bülow, A.; Bégin, P.; Carter, H.; Burns, T., Migration of Volatile Compounds through Stacked Sheets of Paper during Accelerated Ageing. Part II: Variable Temperature Studies. *Restaurator* **2000,** 21, 187-203.

98. Carter, H.; Bégin, P.; Grattan, D., Migration of Volatile Compounds through Stacked Sheets of Paper during Accelerated Ageing. Part I: Acid Migration at 90°C. *Restaurator* **2000,** 21, 77-84.

99. Banik, G.; Stachelberger, H., Phänomene und Ursachen von Farb- und Tintenfraß. In *Wiener Berichte über Naturwissenschaft in der Kunst*, Vendl, A., Pichler, B., Ed. Wien, 1984; Vol. 1, pp 188-213.

100. Daniels, V., Aging of Paper and Pigments Containing Iron and Copper: A Review. In *The Broad Spectrum. Studies in the Materials, Techniques, and Conservation of Color on Paper*, Stratis, H. K., Salvesen, B., Ed. Archetype: London, 2002; pp 116-121.

101. Czepiel, T. P., The Influence of Selected Metal Traces on the Color and Color Stability of Purified Cotton Linters. *Tappi Journal* **1960**, 43, 289-299.
102. Emery, J. A.; Schröder, H. A., Iron-Catalyzed Oxidation of Wood Carbohydrates. *Wood Science and Technology* **1974**, 8, 123-137.
103. Brown, D. G.; Abbot, J., Effects of Metal Ions and Stabilisers on Peroxide Decomposition During Bleaching. *Journal of Wood Chemistry and Technology* **1995**, 15 85-111.
104. Šelih, V. S.; Strlič, M.; Kolar, J.; Pihlar, B., The role of transition metals in oxidative degradation of cellulose. *Polymer Degradation and Stability* **2007**, 92, 1476-1481.
105. Shahani, C.; Hengemihle, F., *The Influence of Copper and Iron on the Permanence of Paper*. American Chemical Society: Washington, D.C., 1986; Vol. 212.
106. Williams, J. C.; Fowler, C. S.; Lyon, M. S.; Merrill, T. I., Metallic Catalysts in the Oxidative Degradation of Paper. In *Preservation of Paper and Textiles of Historic and Artistic Value*, Williams, J. C., Ed. American Chemical Society: Washington D.C., 1977; Vol. 164.
107. Krekel, C., Chemische Struktur historischer Eisengallustinten. In *Tintenfraßschäden und ihre Behandlungsmöglichkeiten*, Banik, G., Weber, H., Ed. Kohlhammer: Stuttgart, 1999; pp 25-36.
108. Kolar, J.; Strlič, M.; Budnar, M.; Malešič, J.; Šelih, V. S.; Simčič, J., Stabilisation of corrosive iron gall inks. *Acta Chimica Slovenica* **2003**, 50, 763-770.
109. Prior, R. L.; Cao, G. In *Antioxidant capacity and polyphenolic components of teas: Implications for altering in vivo antioxidant status*, Proceedings of the Society for Experimental Biology and Medicine 1999; 1999; pp 255-261.
110. Weisburger, J. H., Mechanisms of action of antioxidants as exemplified in vegetables, tomatoes and tea. *Food and Chemical Toxicology* **1999**, 37, 943-948.
111. Strlič, M.; Radovič, T.; Kolar, J.; Pihlar, B., Anti- and prooxidative properties of gallic acid in fenton-type systems. *Journal of Agricultural and Food Chemistry* **2002**, 50, 6313-6317.
112. Neevel, J., Phytate: a Potential Conservation Agent for the Treatment of Ink Corrosion Caused by Irongall Inks. *Restaurator* **1995**, 16, 143-160.
113. Strlič, M.; Kolar, J.; Šelih, V.-S.; Kočar, D.; Pihlar, B., A comparative study of several transition metals in fenton-like reaction systems at circum-neutral pH. *Acta Chimica Slovenica* **2003**, 50, 619-632.
114. Mitchell, C. A., *Inks - Their Composition and Manufacture Including Methods of Examination and Full list of English Patents*. Charles Griffin & Company, LTD: London, 1916.
115. Bergerhoff, G.; Weber, R.; Wunderlich, C. H., Über Eisengallustinten. *Zeitschrift für anorganische und allgemeine Chemie* **1991**, 598/599, 371-376.
116. Zerdoun Bat-Yehouda, M., *Les Encres Noires au Moyen Âge*. Édition du Centre National de la Recherche Scientifique: Paris, 1983.
117. Scott, D. A., Copper and Bronze in Art. Corrosion. Colorants. Conservation. In *Getty Conservation Institute*, Los Angeles, 2002; pp 270-293.
118. Wächter, O., De Viride - destruktive und unschädliche grüne Kupferpigmente in der Buchmalerei. *Biblios* **1981**, 30, 270-284.
119. Neevel, J. In *(Im)possibilities of the phytate treatment*, The Iron Gall Ink Meeting, Newcastle Upon Tyne, 2000; Brown, A. J. E., Ed. The University of Northumbria: Newcastle Upon Tyne, 2000; pp 125-134.
120. Neevel, J. G.; Mensch, C. T. J. In *The behaviour of iron and sulphuric acid during iron gall ink corrosion*, ICOM-CC Lyon 12th Triennial Meeting, 1999b; James & James Ltd. : 1999b.
121. Penders, N. J. M. C.; Havermans, J. B. G. A.; Genuit, W. J. L. In *Emission of volatile organic compounds from objects affected by Iron Gall ink and ink components after*

accelerated ageing, The Iron Gall Ink meeting, Newcastle Upon Tyne, 2000; Brown, A. J. E., Ed. Universtity of Northumbria: Newcastle Upon Tyne, 2000; pp 53-58.

122. Gilbert, A. F.; Pavlovova, E.; Rapson, W. H., Mechanism of Magnesium Retardation of Cellulose Degradation during Oxygen Bleaching. *Tappi Journal* **1973**, 56, 95-99.

123. Yokoyama, T.; Matsumoto, Y.; Meshitsuka, G., Role of peroxide species in carbohydrate degradation during oxygen bleaching. Part III: effect of metal ions on the reaction selectivity between lignin and carbohydrate model compounds. *Journal of Pulp and Paper Science* **1999**, 25, 42-46.

124. Kolar, J.; Šala, M.; Strlič, M.; Šelih, V. S., Stabilisation of Paper Containing Iron Gall Ink with Current Aqueous Processes. *Restaurator* **2005**, 26, 181-189.

125. Barrett, T.; Mosier, C., The role of gelatine in paper permanence. *Journal of the American Institute of Conservation* **1995**, 35, 173-186.

126. Kolbe, G., Gelatine in historical paper production and as inhibiting agent for iron-gall ink corrosion on paper. *Restaurator* **2004**, 25, 26-39.

127. Sixta, H.; Süss, H. U.; Potthast, A.; Schwanninger, M.; Krotscheck, A. W., *Handbook of Pulp*. Wiley-VCH: Weinheim, 2006; Vol. 2, p 609-932.

128. Hofmann, C.; van der Reyden, D.; Baker, M., Comparison and Evaluation of Bleaching Procedures: The Effect of Five Bleaching Methods on the Optical and Mechanical Properties of New and Aged Cotton Linter Paper Before and After Accelerated Aging. *The Book and Paper Group Annual* **1991**, 10, no pages.

129. Trosper Schaeffer, T.; Baker, M. T.; Blyth-Hill, V.; van der Reyden, D., Effects of Aqueous Light Bleaching on the Subsequent Aging of Paper. *Journal of the American Institute of Conservation* **1992**, 31, 289-311.

130. Bicchieri, M.; Brusa, P., The Bleaching of Paper by Reduction with Borane Tert-Butylamine Complex. *Restaurator* **1997**, 18, 1-11.

131. Bicchieri, M.; Bella, M., Semetilli, F., A Quantitative Measure of Borane Tert-Butylamine Complex Effectiveness in Carbonyl Reduction of Aged Papers. *Restaurator* **1999**, 20, 22-29.

132. Pedersoli Jr., J. L.; Ligterink, F. J., Spectroscopic characterization of the fluorescence of paper at the wet-dry interface. *Restaurator* **2001**, 22, 133-145.

133. Hutchins, J. K., Water-Stained Cellulosics: A literature review. *Journal of the american institute of conservation* **1983**, 22, 57-61.

134. Eusman, E., Tideline Formation in Paper Objects: Cellulose Degradation at the Wet-Dry Boundary. *Studies in the History of Art 51. Monograph Series II: Conservation Research, Washington, DC: National Gallery of Art* **1995**, 11-27.

135. Whitmore, P. M.; Bogaard, J., Determination of the Cellulose Scission Route in the Hydrolytic and Oxidative Degradation of Paper. *Restaurator* **1994**, 15, 26-45.

136. Blüher, A.; Vogelsanger, B., Mass deacidification of paper. *Chimia* **2001**, 55, 981-989.

137. Bansa, H., Aqueous Deacidification - with Calcium or with Magnesium? *Restaurator* **1998**, 19, 1-40.

138. Bukovský, V., Yellowing of newspaper after deacidification with methyl magnesium carbonate. *Restaurator* **1997**, 18, 25-38.

139. Kolar, J., Mechanism of autoxidative degradation of cellulose. *Restaurator* **1997**, 18, 163-176.

140. Strlič, M.; Kolar, J.; Malesič, J.; Kočar, D.; Šelih, V.-S.; Pihlar, B.; Haillant, O.; Pedersoli, J. L.; Scholten, S.; Rychlý, J.; Rychlá, L.; Fromageot, D.; Lemaire, J., Stabilisation Strategies. In *Ageing and Stabilisation of Paper*, Strlič, M., Kolar, J., Ed. National and University Library: Ljubljana, 2005; pp 181-198.

141. Hofmann, R.; Wiesner, H. J., *Bestandserhaltung in Archiven und Bibliotheken*. 1 ed.; DIN Deutsches Institut für Normung e.V.: Berlin, 2007; p 1-268.

142. Blüher, A. In *The alkaline reserve - a key element in paper deacidification*, Save Paper!, Bern, 2006; Blüher, A., Grossenbacher, G., Banik, G., Ed. Swiss National Library: Bern, 2006; pp 191-206.

143. Banik, G., Mass Deacidification Technology in Germany and its Quality Control. *Restaurator* **2005**, 26, 63-75.

144. Banik, G.; Mairinger, F.; Stachelberger, H., Erscheinungen und Probleme des Kupferfraßes in der Buchmalerei. *Restaurator* **1981**, 1-2, 71-93.

145. Banik, G.; Stachelberger, H., Identifizierung von Abbauprodukten der Zellulose in mit Kupferpigmenten illuminierten Graphiken. *Monatshefte für Chemie* **1982**, 113, 845-848.

146. Fiedler, A., Wall Decorations on Paper in Baroque Castles in Austria Part II: Chinese Wallpapers and Their Restoration. In *Conservation within Historic Buildings*, Brommelle, N. S.; Thomson, G.; Smith, P., Eds. ICC: Vienna, 1980.

147. Kolar, J.; Strlič, M. In *Stabilisation of Ink Corrosion*, The Iron Gall Ink Meeting, Newcastle Upon Tyne, 2000; Brown, J. E., Ed. University of Northumbria: Newcastle Upon Tyne, 2000.

148. Kühn, H., Grünspan und seine Verwendung in der Malerei. *Farbe und Lacke* **1964**, 70, 703-711.

149. Schweizer, F.; Mühlethaler, B., Einige grüne und blaue Kupferpigmente: Herstellung und Identifikation. *Farbe und Lacke* **1968**, 74, 1159-1173.

150. Shahani, C. J.; Hengemihle, F. H. *Effect of Some Deacidification Agents on Copper-Catalyzed Degradation of Paper*; Library of Congress: Washington D.C., 1995; p 13.

151. Fischer, M. Polymeranaloge Carbanilierung von Cellulose. Beiträge zur Methodenentwicklung und Untersuchung von Depolymerisationsprozessen. Technische Universität Dresden, Dresden, 2004.

152. Savitzky, A.; Golay, M. J. E., Smoothing and differentiation of data by simplified least squares procedures. *Analytical Chemistry* **1964**, 36, 1627-1639.

153. Prohaska, T.; Latkoczy, C.; Schultheis, G.; Teschler-Nicola, M.; Stingeder, G., Investigation of Sr isotope ratios in prehistoric human bones and teeth using laser ablation ICP-MS and ICP-MS after Rb/Sr separation. *Journal of Analytical Atomic Spectrometry* **2002**, 17, 887-891.

154. Kolar, J.; Stolfa, A.; Strlic, M.; Pompe, M.; Pihlar, B.; Budnar, M.; Simcic, J.; Reissland, B., Historical Iron Gall Ink Containing Documents - Properties Affecting their Condition. *Analytica Chimica Acta* **2006**, 555, 167-174.

155. Bicchieri, M.; Pepa, S., The Degradation of Cellulose with Ferric and Cupric Ions in a Low-acid Medium. *Restaurator* **1996**, 17, 165-183.

156. Ranby, B. G., "Weak Links" in Polysaccharide Chains a srelated to Modified Groups. *Journal of Polymer Sciences* **1961**, 53, 131-140.

157. Henniges, U.; Schröter, K., Copper-corroded Wallpapers. In-situ treatment with non-aqueous magnesium compounds? *PapierRestaurierung* **2005**, 6, 25-32.

158. Wouters, J., Codex 801. In KIK: Brussels, 2004.

159. Wappenschmidt, F., *Chinesische Tapeten für Europa. Vom Rollbild zur Bildtapete*. Deutscher Verlag für Kunstwissenschaften: Berlin, 1989.

160. Hahn, O.; Malzer, W.; Kanngiesser, B.; Beckhoff, B., Characterization of iron-gall inks in historical manuscripts and music compositions using x-ray fluorescence spectrometry. *X-Ray Spectrometry* **2004**, 33, 234-239.

161. Quillet, V.; Rémazeilles, C.; Bernard, J.; Buisson, N.; Bouvet, S.; Nguyen, T. P.; Éveno, M. In *Dégradation du Papier Provoquées par L'Emploi de Verts de Cuivre: Test sur des Éprouvettes de Laboratoire d'un Traitement de Restauration a Base d'Acide Phytique et de Carbonate de Calcium*, La Conservation a l'Ère Numérique, Paris, 2002; ARSAG: Paris, 2002; pp 218-232.

162. Banik, G.; Stachelberger, H.; Mairinger, F.; Vendl, A.; Ponahlo, J., Analytical investigations of the problem of "Kupferfraß" in illuminated manuscripts. *Mikrochimica Acta* **1981**, 75, 49-55.
163. Daniels, V., Oxidative damage and the preservation of organic artefacts. *Free Radical Research Communications* **1989**, 5, 213-220.
164. Rosenau, T.; Potthast, A.; Adorjan, I.; Hofinger, A.; Sixta, H.; Firgo, H.; Kosma, P., Cellulose solutions in N-methylmorpholine-N-oxide (NMMO) - degradation processes and stabilizers. *Cellulose* **2002**, 9, 283-291.
165. Vyprachticky, D.; Pospisil, J.; Sedlar, J., Antioxidants and Stabilizers. Part 110. Photo-oxidation of the model system heptane-2-butylanthraquinone in the presence of a hindered piperidine and a phenolic antioxidant. *Polymer Degradation and Stability* **1990**, 27, 49-63.
166. Pospisil, J., Chemical and photochemical behaviour of phenolic antioxidants in polymer stabilization - a state of the art report. Part I. *Polymer Degradation and Stability* **1993**, 40, 217-232.
167. Baty, J.; Barret, T., Gelatin Size as a pH and Moisture Buffer in Paper. *Journal of the American Institute of Conservation* **2007** 45, 105-121.
168. Porck, H. J. *Mass Deacidification - an Update of Possibilites and Limitations*; European Commission on Preservation and Access (ECPA) and Commission on Preservation and Access (CPA): Amsterdam and Washington, 1996.
169. Knill, Degradation of Cellulose under Alkaline Conditions. *Carbohydrate Polymers* **2003**, 51, 281-300.
170. Steuri, B.; Graf, C.; Streiff, D.; Jauslin, J. F. Qualitätsstandards für die Neutralisierung der Materialien des Schweizerischen Bundesarchivs und des Bundesamtes für Kulur, Schweizerische Landesbibliothek nach dem PaperSave Swiss-Verfahren. http://www.nb.admin.ch/slb/slb_professionnel/erhalten/00699/01491/01492/index.html?lang=de
171. Sjöholm, E. Characterisation of Kraft Pulps by Size-Exclusion Chromatography and Kraft Lignin Samples by Capillary Zone Electrophoresis. KTH, Stockholm, 1999.
172. Potthast, A., Chemistry of Kraft Cooking. In *Handbook of Pulp*, Sixta, H., Ed. Wiley-VCH: Weinheim, 2006; Vol. 1, pp 164-185.
173. Henniges, U.; Kloser, E.; Patel, A.; Potthast, A.; Kosma, P.; Fischer, M.; Fischer, K.; Rosenau, T., Studies on DMSO-containing carbanilation mixtures: Chemistry, oxidations and cellulose integrity. *Cellulose* **2007**, 14, 497-511.
174. Sundholm, F.; Tahvanainen, M., Preparation of cellulose samples for size-exclusion chromatography analyses in studies of paper degradation. *Journal of Chromatography A* **2003**, 1008, 129-134.
175. Kim, K. H.; Hong, J., Supercritical CO2 pretreatment of lignocellulose enhances enzymatic cellulose hydrolysis. *Bioresource Technology* **2001**, 77, 139-144.
176. Zheng, Y.; Lin, H.-M.; Tsao, G. T., Pretreatment for Cellulose Hydrolysis by Carbon Dioxide Explosion. *Biotechnology Progress* **1998**, 14, 890-896.
177. Chirat, C.; Viardin, M. T.; Lachenad, D., Use of a reducing stage to avoid degradation of softwood kraft pulp after ozone bleaching. *Paperi ja Puu/Paper and Timber* **1994**, 76, 417-422.
178. Bouchard, J.; Méthot, M.; Jordan, B., The effects of ionizing radiation on the cellulose of woodfree paper. *Cellulose* **2006**, 13, 601-610.
179. Potthast, A.; Kostic, M.; Schiehser, S.; Kosma, P.; Rosenau, T., Studies on oxidative modifications of cellulose in the periodate system: Molecular weight distribution and carbonyl group profiles. *Holzforschung* **2007**, 61, 662-667.
180. Berggren, R. Cellulose Degradation in Pulp Fibres Studied as Changes in Molar Mass Distributions. KTH, Stockholm, 2003.

181. Yanagisawa, M.; Shibata, I.; Isogai, A., SEC-MALLS analysis of cellulose using LiCl/1,3-dimethyl-2-imidazolidinone as an eluant. *Cellulose* **2004**, 11, 169-179.

182. Isogai, A.; Kato, Y., Preparation of polyuronic acid from cellulose by TEMPO-mediated oxidation. *Cellulose* **1998**, 5, 153-164.

183. Rodrigues, J.; Alves, A.; Pereira, H.; Da Silva Perez, D.; Chantre, G.; Schwanninger, M., NIR PLSR results obtained by calibration with noisy, low-precision reference values: Are the results acceptable? *Holzforschung* **2006**, 60, 402-408.

184. Antti, H.; Alexandersson, D.; Sjöström, M.; Wallbäcks, L., Detection of kappa number distributions in kraft pulps using NIR spectroscopy and multivariate calibration. *Tappi Journal* **2000**, 83, 102-108.

185. Antti, H.; Sjöström, M.; Wallbäcks, L., Multivariate calibration models using NIR spectroscopy on pulp and paper industrial applications. *Journal of Chemometrics* **1996**, 10, 591-603.

186. Brunner, M.; Eugster, R.; Trenka, E.; Bergamin-Strotz, L., FT-NIR spectroscopy and wood identification. *Holzforschung* **1996**, 50, 130-134.

187. Muños Vinas, S., *Contemporary Theory of Conservation*. first ed.; Elsevier Butterworth-Heinemann: Oxford, 2005.

188. McCrady, E., The Great Cotton-Rag Myth. *Alkaline Paper Advocate* **1992**, 5, no pages.

Die VDM Verlagsservicegesellschaft sucht für wissenschaftliche Verlage abgeschlossene und herausragende

Dissertationen, Habilitationen, Diplomarbeiten, Master Theses, Magisterarbeiten usw.

für die kostenlose Publikation als Fachbuch.

Sie verfügen über eine Arbeit, die hohen inhaltlichen und formalen Ansprüchen genügt, und haben Interesse an einer honorarvergüteten Publikation?

Dann senden Sie bitte erste Informationen über sich und Ihre Arbeit per Email an *info@vdm-vsg.de*.

Sie erhalten kurzfristig unser Feedback!

VDM Verlagsservicegesellschaft mbH
Dudweiler Landstr. 99
D - 66123 Saarbrücken

Telefon +49 681 3720 174
Fax   +49 681 3720 1749

www.vdm-vsg.de

Die VDM Verlagsservicegesellschaft mbH vertritt

Printed by Books on Demand GmbH, Norderstedt / Germany